Learning Systems

Springer
London
Berlin
Heidelberg
New York
Barcelona
Budapest
Hong Kong
Milan
Paris
Santa Clara
Singapore
Tokyo

Eduard Aved'yan

LEARNING SYSTEMS

Edited by J. Mason and P.C. Parks

With 15 Figures

 Springer

Eduard Aved'yan, PhD

Institute of Control Sciences
Laboratory 07
Profsoyuznaja 65
Moscow 177806
Russia

ISBN-13: 978-3-540-19996-0   e-ISBN-13:978-1-4471-3089-5

DOI: 10.1007/ 978-1-4471-3089-5

British Library Cataloguing in Publication Data
Aved'yan, Eduard
   Learning Systems
   I. Title
   006.31

Library of Congress Cataloging-in-Publication Data
A catalog record for this book is available from the Library of Congress

© Springer-Verlag London Limited 1995
Softcover reprint of the hardcover 1st edition 1995

Typesetting: Camera ready by authors

69/3830-543210 Printed on acid-free paper

This book is dedicated to the memory
of the late Professor Parks

Foreword

Eduard Aved'yan is a senior scientific researcher at the Institute of Control Sciences in Moscow. Born in Armavir, he graduated from the Erevan Polytechnical Institute with a degree in Electrical Engineering and went on to receive his Candidate of Science (equivalent to a Ph.D) in 1972 from the Institute of Control Sciences. He has held invited positions as guest researcher at the Technical University in Ilmenau, Germany and more recently at the University of Darmstadt. In late 1993 he was a guest professor at the Polytechnical Institute of Milan. His current areas of interest are mathematical methods and their application to neural networks, digital filters, digital image processing and adaptive control and he has published widely in international journals. Fluent in English, German, Armenian and Russian, Dr. Aved'yan lives in Moscow with his wife and child.

The lecture course transcribed in this book was first given by Dr. Aved'yan at the Polytechnical Institute of Milan. During a visit to Moscow by Professor Parks and Julian Mason in October 1994 the contents of the lectures were the subject of a series of seminars and technical meetings held at the Institute. It was decided to publish the material to make it available to a wider audience.

Preface

A learning system can be loosely defined as a system which can adapt its behaviour to become more effective at a particular task or set of tasks. Generally learning systems are taught using a set of training examples presented by an expert or teacher, and are called supervised learning systems. In addition there are autonomous or unsupervised learning systems which change their behaviour automatically on the basis of natural clusters within the training data. A learning system consists of two components: an architecture with a set of variable parameters and an algorithm to update these parameters in order to improve the performance of the overall system. The system must therefore have some memory to retain the current values of its parameters. The learning algorithm needs a measure of the accuracy (or error) at the system output to be able to determine how the parameters should be changed to improve the system, this measure is called the performance index.

Learning systems are useful in many fields, one of the major areas is in control and system identification. They can be used to form automatically a model of a system which can be employed in fault diagnosis or to develop a controller in an indirect control scheme. Alternatively the parameters of a controller can be updated automatically to form a direct adaptive control system. They also have applications in cognitive science where they can be used as pattern recognition devices or as classifiers.

This lecture series covers the major aspects of learning systems: system architecture, choice of performance index and methods of measuring error, and the major learning algorithms are explained in full detail including proofs of convergence. Artificial neural networks are an important class of learning systems which have been subject to rapidly increasing popularity, they are covered in detail in chapters 6 to 9. Where appropriate, examples are given to demonstrate the practical use of techniques developed in the text. Chapters 8 and 9 discuss system identification and control using multi-layer networks and also using CMAC.

Table of Contents

1 Introduction to Learning Systems

1.1 Systems, Memory

Consider a system S with an input vector u and an output vector y. The system is described by the operator T, so that $y = T(u)$. This system may be a system of linear differential or difference equations, a system of non-linear equations, a system of linear algebraic equations, and so on. Such kinds of systems may describe many physical processes.

Suppose for example that we have a system of linear difference equations:

$$x(n) = Ax(n-1) + Bu(n-1),$$ (1.1)
$$y(n) = Cx(n),$$

where $x \in R^{N_x}, u \in R^{N_u}$ and $y \in R^{N_y}$ are the state vector, the input vector and the output vector of the system respectively; $A \in R^{N_x \cdot N_x}$, $B \in R^{N_x \cdot N_u}$ and $C \in R^{N_y \cdot N_x}$ are matrices of appropriate dimensions; $n = 1, 2, ...$ is the discrete time. In this case the operator T, that describes the connection between vectors u and y can be written as

$$T = C(I - Az^{-1})^{-1} Bz^{-1} .$$ (1.2)

Here z^{-1} is the shift operator: $z^{-1}y(n) = y(n-1)$, $(I - Az^{-1})^{-1}$ is the inverse of the matrix $(I - Az^{-1})$, I is a unit matrix.

Suppose that the initial state vector $x(0)$ of system (1.1) is fixed; then the fixed matrices A, B and C and the fixed input function $u(n)$ yield a fixed output function $y(n)$. For example, if $x \in R^1, u \in R^1$, $b = c = 1; u(n) = 1$, $n \geq 1$ and $x(0)=0$ then the output of the system, for different values of parameter a, has the following behaviour:

a) if $a= 0.5$ then $y(n) = 2(1-(0.5)^n)$ is varying exponentially,

b) if $a=-0.5$ then $y(n) = 2(1-(-0.5)^n)$ is oscillatory,

c) if $a=1$ then $y(n) = n$ is varying linearly.

Why are the motions of the same simple system S^1 with the fixed input function $u(n)$ so different? The values of parameter a in the three cases are different. The parameters a, b and c are playing the role of the system memory $w = (a, b, c)$: hence we will identify the system parameters with the system memory, and the connection between input and output of the system will be described by the equation

$$y = T(u;w) .$$
(1.3)

1.2 Performance Index

After the introduction to the concept of a system we can give the definition of a learning system:

A learning system is a system that can change its structure and parameters (i.e. the memory) so that the output signal y(n) approaches the desired response for all possible input signals. The desired response will be called the teacher's instruction.

It follows from the definition that a learning system has a memory and that the memory is changeable.

The closeness of the system output signal $y(n)$ to the teachers instruction $y^*(n)$ can be described by a convex function $Q(y^*(n), y(n))$, often given as a function of the system error $\varepsilon(n) = y^*(n) - y(n) = \varepsilon(n;w)$:

$$Q(y^*(n), y(n)) = Q(\varepsilon(n;w)) .$$
(1.4)

The closeness of the system output signal to the teacher's instruction must hold for all possible input signals $u(n)$.

1.2.1 Random Input

If the input signal (or the pattern) $u(n)$ is a random one with probability density function $p(u)$, then for all possible input signals the closeness of the system output to the teacher's instruction can be given by the performance index (PI) $J(w)$, which is an average of $Q(\varepsilon(n;w))$ over U:

$$J(w) = \int_U Q(y^*, y)p(u)du \,. \tag{1.5}$$

Here it is assumed that $y^*(n) = y^*(n;u)$ and $y(n) = y(n;u,w)$. If we don't know the probability density function $p(u)$ *a priori*, then we cannot calculate the performance index $J(w)$, but it is possible to estimate it in the following way: Let $u(m), m = \overline{1,n}$ be the observed input signals. Then an empirical estimate of the probability density function $p(u)$ is

$$\hat{p}(u) = \frac{1}{n}\sum_{m=1}^{n}\delta(u - u(m)) \,, \tag{1.6}$$

where $\delta(u)$ represents a multi-dimensional δ-function: $\delta(u) = \infty$ if $u=0$; $\delta(u) = 0$, if $u \neq 0$ and $\int \delta(u)du = 1$. Then the estimate of the performance index $J(w)$ can be given in the following form:

$$\hat{J}(w) = \int_U Q(y^*, y)\hat{p}(u)du = \frac{1}{n}\sum_{m=1}^{n}Q(y^*(m), y(m)) \,. \tag{1.7}$$

Another approach for the construction of the PI is based on the generalised maximum likelihood method: the PI is constructed as *a posteriori* probability density function for the memory w. It is assumed that a sequence of the teacher's instructions $y^*(1), y^*(2), \ldots y^*(n)$ is available and the teacher's instructions have the following form

$$y^*(n) = T(u(n), w^*) + \xi(n), \tag{1.8}$$

where $T(u(n), w^*)$ is an output of the system and $T(u; w^*)$ is an operator of the system, the memory w^* is unknown, $\xi(n)$ is a random error with probability density function $p_\xi(\xi)$ and all errors $\xi(m), m = \overline{1,n}$ are independent.

If the memory w is random and we know its *a priori* density function $p_{a\ pri}(w)$, then we can write its *a posteriori* probability density function in the following form [1]:

$$p_{a\ post}(w) = p_{a\ pri}(w) \prod_{m=1}^{n} p_\xi(y^*(m) - T(u(m), w^*)).$$ (1.9)

When we don't know the a priori probability density function $p_{a\ pri}(w)$, we can write the *a posteriori* density function as

$$p_{a\ post}(w) = \prod_{m=1}^{n} p_\xi(y^*(m) - T(u(m), w^*)).$$ (1.10)

The maximum likelihood *PI* is constructed by taking the logarithm of the *a posteriori* probability density function (1.9) or (1.10) :

$$J(w) = -\ln p_{a\ post}(w) =$$

$$-\ln p_{a\ pri}(w) - \sum_{m=1}^{n} \ln p_\xi(y^*(m) - T(u(m), w^*))$$ (1.11)

or

$$J(w) = -\sum_{m=1}^{n} \ln p_\xi(y^*(m) - T(u(m), w^*))$$ (1.12)

respectively. It follows from expressions (1.7), (1.11) and (1.12) that the *PI* is usually a function of the discrete time n (the number of the pattern presentations).

1.2.2 Deterministic Input

In the particular case when the number P of input vectors, or patterns, is finite, a common *PI* is:

$$J(w) = \frac{1}{P} \sum_{m=1}^{P} Q(\varepsilon(m; w)).$$ (1.13)

Often $Q(\varepsilon(m))$ is the (weighted) sum of squares of the components of the vector error

$$Q(\varepsilon(n)) = \sum_{i=1}^{N_y} \alpha_i \varepsilon_i^2(n),$$ (1.14)

where $\alpha_i \geq 0$ are weights characterising the importance of component $\varepsilon_i(n)$ of the error.

1.3 Learning Algorithms

The quality of the learning system is measured by a performance index $J(w)$. The goal of the learning system is to find the extremum of the performance index, i.e. the minimum or maximum, depending on the particular problem. This objective is accomplished by changing the system's memory w with suitable learning algorithms.

There are two different approaches for obtaining the learning algorithms.

1. If the performance index $J(w)$ is not a function of n, (see expressions (1.5) and (1.13)), then we can use various optimisation algorithms for finding the optimal value of the memory w^*: the gradient method, Newton's method, the search method, the Monte-Carlo method and so on, depending on what is known about the performance index $J(w)$ [2], [3]. For example, if the PI is differentiable with a unique minimum, and it is possible to calculate the gradient of the performance index $J(w)$, then its minimum can be reached with the gradient algorithm:

$$w(n) = w(n-1) - \gamma(n)\nabla_w J(w(n-1)), \ n = 1, 2, \ldots$$ (1.15)

Here $\gamma(n) \geq 0$ is the step size, $\nabla_w J(w) = \left(\dfrac{\partial J(w)}{\partial w_1}, \ldots, \dfrac{\partial J(w)}{\partial w_N} \right)^T$ is the gradient of the PI (a column-vector, consisting of N partial derivatives $\partial J(w)/\partial w_i, \ i = \overline{1, N}$ of the PI with respect to all the components of the vector memory w).

2. If the PI is a function of n, then it is still possible to use the approach described above for seeking the extremum of the PI. However, this approach is not convenient, because in this case we need a large memory and a large computational effort for each memory correction step. For example, in the case (1.7) the gradient algorithm has the form

$$w(n) = w(n-1) - \gamma(n)\frac{1}{n}\sum_{m=1}^{n}\nabla_w Q(m;w(n-1)), \, n = 1, 2, \ldots \qquad (1.16)$$

where the initial value of the memory $w(0) = w_0$. Instead of algorithm (1.16) for finding the optimal value w^* it is more convenient to use the stochastic approximation algorithm, in which the gradient $\nabla_w J(w)$ of the function $J(w)$ at the point $w(n-1)$ or its estimate $\frac{1}{n}\sum_{m=1}^{n}\nabla_w Q(m;w(n-1))$ is replaced by the sample value $\nabla_w Q(n;w(n-1))$, to yield the following algorithm:

$$w(n) = w(n-1) - \gamma(n)\nabla_w Q(n;w(n-1)), \, n = 1, 2, \ldots \qquad (1.17)$$

We have to impose certain specific constraints on the sequence $\gamma(n)$ in order to ensure convergence for algorithm (1.17). For example, in many cases $\gamma(n) = 1/n$. It is possible to explain in the following way why we use only the last member of the sum instead of the whole sum in algorithm (1.17).

At step n-1 we have to solve the equation $\frac{1}{n-1}\sum_{m=1}^{n-1}\nabla_w Q(m;w) = 0$. If $w(n$-$1)$ is the (unknown) solution of this equation, then in algorithm (1.16) the sum $\sum_{m=1}^{n}\nabla_w Q(m;w(n-1))$ is equal only to the last member in the sum and algorithm (1.16) turns into algorithm (1.17). We shall discuss the details of the algorithms in the following chapters. A comprehensive introduction to adaptive and learning systems can be found in [4], [5].

1.4 Some Examples of Learning Systems

1.4.1 The Learning Linear Combiner

A basic system identification tool is an adaptive linear combiner [6] described by the following equation

$$y(n) = \sum_{j=1}^{N} w_j u_j(n) = w^T u(n). \qquad (1.18)$$

In Eq.(1.18) the components of the input vector $u(n) = (u_1(n),...,u_N(n))^T$ at time n are weighted by a set of coefficients $w = (w_1,...,w_N)^T$ (memory of the combiner), producing a linear output $y(n)$. During the learning process, the input signals, or patterns, and the desired outputs are presented to the linear combiner. There are a lot of algorithms which can automatically adjust the weights w so that the output signal $y(n)$ will be as close as possible to the desired signal (teacher's instruction) $y^*(n)$, for example the simple recursive least squares algorithm, the weighted recursive least squares algorithm (algorithm with a forgetting factor), the Kaczmarz algorithm, the modified Kaczmarz algorithm, row-action algorithms and Kalman's algorithm. One of the most popular algorithms for adapting the weights w is the simple least mean squares algorithm, for which the performance index has the following form

$$J(w) = \sum_{m=1}^{n} (y^*(m) - w^T u(n))^2.$$

The choice of the algorithm depends on the mathematical model of the teacher's instruction, or on the *a priori* information. The details will be discussed in the following chapters.

1.4.2 Neurons of Higher Order

These learning elements consist of an adaptive linear combiner cascaded with a hard-limiting quantiser (which is used to produce binary outputs (+1,-1) or (+1,0)) or with a sigmoidal non-linearity [6]. A typical sigmoidal function is the hyperbolic tangent:

$$y = \tanh(net) = \frac{1 - e^{-2net}}{1 + e^{-2net}}.$$ (1.19)

For first order neurons the function *net* in Eq.(1.19) is an output of the linear combiner described by Eq.(1.18). For a neuron of higher, say of order K, the function *net* in Eq.(1.19) is the polynomial function of the components of the input vector:

$$net = \sum_{i_1,...,i_k} u_{i_1} u_{i_2} ... u_{i_k} w_{i_1,...,i_k}.$$ (1.20)

In Eq.(1.20) the $u_{i_j}, j = \overline{1,k}, i_j = \overline{1,N}$ are the components of the input vector $u(n) = (u_1(n),...,u_N(n))^T$. Here it is assumed that $u_1 = 1$. For example, the second order neuron is described by the equation

$$y = \tanh(net) = \frac{1 - e^{-2net}}{1 + e^{-2net}} , \tag{1.21}$$

$$net = w_1 + \sum_{i=2}^{N} w_i u_i + \sum_{i=2, j=2; j \geq i}^{N} u_i u_j w_{i,j} . \tag{1.22}$$

The optimal memory value of this neuron can be obtained using the gradient method. We shall discuss the details in the chapter devoted to multilayer neural networks.

1.4.3 The Learning Non-Linear Transformer

In this case the teacher's instructions $y^*(n)$ are some non-linear transformations $\varphi(u)$ of the *a priori* unknown structure of the input signal $u(n)$. Here we shall introduce the class of piecewise-linear transformers. Other types of transformations are obtained using a segment of the Fourier or Taylor series as non-linear transformer.

For the sake of simplicity we will consider here the one dimensional case, $y, y^*, u \in R^1$. The approximation of the function $\varphi(u)$ is given by expression [7]:

$$y = \sum_{i=1}^{N} (w_i^0 + u w_i^1) 1(u, u_i, u_{i-1}). \tag{1.23}$$

In Eq. 23 the function $1(u, u_i, u_{i-1})$ is a window type function

$$1(u, u_i, u_{i-1}) = \begin{cases} 1, & \text{if } u_{i-1} < u \leq u_i \\ 0 & \text{otherwise} \end{cases} \tag{1.24}$$

and $u_i, i = \overline{0, N}$ are the prescribed points lying on the U axis. The weights $w_i^0, w_i^1, i = \overline{1, N}$ constitute the memory of the system, w. If the performance index is quadratic, then we can use the recursive least squares algorithm for the learning, and the learning process for each pair $w_i^0, w_i^1, i = \overline{1, N}$ is independent.

1.4.4 Linear and Non-Linear Learning Filters

Let us suppose that the signal $s(n)$ is transmitted through a channel with the transfer function $W(z)$, and that the received signal at the end of the channel is the transformed signal $W(z)s(n)$ plus some noise signal $\xi(n)$:

$$u(n) = W(z)s(n) + \xi(n). \tag{1.25}$$

We can extract the signal $s(n)$ from the observations $u(n)$ using a Kolmogorov-Wiener filter, if the signal $s(n)$ and the noise $\xi(n)$ are stationary random processes. For example, a non-realisable filter that minimises the error variance, is given by the following transfer function [8]:

$$K_{opt}(z) = \frac{W(z^{-1})S_s(z,z^{-1})}{W(z^{-1})W(z)S_s(z,z^{-1}) + S_\xi(z,z^{-1})}, \tag{1.26}$$

where $S_s(z,z^{-1})$ and $S_\xi(z,z^{-1})$ are spectral density functions of the signal and the noise. It follows from Eq.(1.26) that for the construction of the optimal filter it is necessary to know the transfer function of the channel and the spectral densities of the signal and noise. Instead of the classical Kolmogorov-Wiener filter (1.26) it is possible to construct a learning filter with *a priori* given structure. In this case it is not necessary to know the spectral densities and the transfer function of the channel. The learning approach can be especially powerful in the case when the channel has slowly time varying characteristics and it is not possible to use the Kolmogorov-Wiener theory.

In the learning case the process $u(n)$ feeds N linear filters with linearly independent impulse functions $k_\nu(n)$, $\nu = \overline{1,N}$, (e.g., time delay elements). Each output signal $\varphi_\nu(n) = K_\nu(z)u(n)$ is multiplied by a weight w_ν, $\nu = \overline{1,N}$ and the sum of such signals is the output of the filter

$$y(n) = \sum_{\nu=1}^{N} w_\nu K_\nu(z)u(n). \tag{1.27}$$

When the learning filter is made of delay elements [9], then the output $y(n)$ can be written as

$$y(n) = \sum_{\nu=1}^{N} w_\nu u(n-\nu). \tag{1.28}$$

The teacher's instruction for this problem is the transmitted signal $s(n)$, while the weights are playing the role of the memory of the learning filter. The following performance indices can be used

$$J(w) = E\left\{\left(s(n) - y(n)\right)^2\right\} \qquad (1.29)$$

or

$$J(w) = \frac{1}{n} \sum_{m=1}^{n} \left(s(m) - y(m)\right)^2 \qquad (1.30)$$

It follows from Eqs.(1.29), (1.30), that for the learning filter it is necessary to know the transmitted signal $s(n)$. In the case (1.30) the learning algorithm is the well known recursive least squares algorithm.

1.4.5 Learning control systems

Two distinct approaches have been used in adaptive control theory: direct adaptive control and indirect adaptive control. In indirect adaptive control the plant parameters are unknown and are estimated from measurements at each time step. The controller parameters are calculated assuming that the estimates represent the true values of the plant parameters. In direct adaptive control, the parameters of the controller are directly adjusted to minimise some performance index. The second case is a typical setting for learning systems.

At present, very interesting results for linear time-invariant plants with unknown parameters are known because the controller in this case is also linear and its structure can be easily found. Here the learning system is a linear one and has a structure similar to that of the plant to be controlled.

For non-linear plants it is clear that the structure of the controller is also non-linear, but only in particular cases it is possible to find the structure in a simple form. This is the main reason why multilayer neural networks are widely used in learning control systems, in fact, they can approximate any non-linear controller function with desired accuracy without knowledge of the structure of the controller[10].

1.4.6 The Learning Multilayer Neural Network

Artificial Neural Networks (NNs) are an alternative to the conventional Von Neumann stored program method of computing. Usually to solve a particular problem using a computer it is necessary to know the algorithm required and then to code this in the form of a program. The computer can then execute this program at high speed and accuracy making it ideal for solving many problems. However, for some applications the algorithm is not known. For instance although humans have little difficulty distinguishing one face from another or recognising an object regardless of its scale or orientation, it is very difficult to describe exactly how we do it in terms of an algorithm that can be used to program a computer to perform the same task.

The exact method by which the human brain operates is unknown at present. Investigations have determined that it consists of a highly interconnected network of neurons. These are special cells which appear to be able to perform simple operations on incoming nerve impulses and to produce an output according to the result. Animal and human brains excel at the tasks computers find difficult, therefore NNs were developed to attempt to model the operation of real brains. As in real brains, NNs consist of an array of interconnected processing elements called nodes or neurons. The neurons in a brain are connected via synapses which can vary their conductivity, it is thought that learning takes place by adaptation of the conductivity or strengths of the synapses. The current node (neuron) models are very crude analogies of the real item, generally they perform a very simple function such as a weighted sum of the inputs which is passed through some activation function e.g. a threshold. The weights of the summation are analogous to the synapses of a real brain, and learning algorithms train the network by varying these weights. The interconnectivity and overall size of NNs are both drastically less than their biological counterparts. The essential difference between NNs and conventional computing methods is that instead of having a stored program, NNs learn by experience. NNs have a set of parameters (the weights) which are adapted by a learning algorithm according to some performance index and so are an example of a learning system.

The manner in which the nodes are interconnected is referred to as the architecture or topology of the network. NNs can be divided into two main classes on the basis of their architecture. Recurrent networks, e.g. Hopfield networks, have feedback loops where the input to a particular node may be affected by that node's output (usually after having been fed through several other nodes). Feedforward networks, as their name suggests, do not have feedback loops, they tend to be arranged in a layer structure. One layer forms the input to the network and another the output, between them can be any number of layers with the common feature that each layer takes its inputs only from preceding (closer to the input) layers.

There are many different training algorithms to choose from. Again these can broadly be divided into two classes: supervised and unsupervised. Supervised regimes provide the network with a set of example input pattern and desired output combinations, the learning algorithm adapts the parameters of the network to minimise the error between the network output and the desired output for all patterns in the training set. Unsupervised algorithms, for example Kohonen networks, attempt to find natural `clusters' (areas of high density) within the input space, as such they attempt to generate their own performance index on the basis of the distribution of the input data.

Being designed according to the principles of their biological analogues, artificial neural networks (NN) are able to solve a wide range of problems in pattern recognition [11], identification [12], control of complex dynamical systems [13],[14], robot control [15], etc.

Let us introduce an M-layered feed-forward NN. Such networks are characterised as follows. The outputs of the m-th layer neurons are fed to the inputs of a part of the neurons of the ($m+1$)-th layer only, an external signal u is fed to the inputs of neurons of the first layer, and the outputs of the neurons of the **M-th** layer form the vector of the overall output of the network. Every v-th neuron of the μ-th layer ((μ,v) – th neuron) is described, for example by the mathematical model (1.21), (1.22).

The optimal memory value of the NN can be computed by the gradient method. The learning process of multilayer neural networks is carried out by comparison of the network's outputs (i.e., the outputs of the neurons of the last layer) with the corresponding teacher's commands. Information about the ideal outputs of the neurons of the hidden layers is absent. Despite this, the knowledge of the structure of an NN allows us to calculate correcting signals not only for the neurons of the last layer but also for all the other neurons. At present one of the most widely used methods for computing correcting signals in neural networks is the back-propagation method [16], [17], [18] which effectively realises a recursive method for computing the gradient of the functional under minimisation. We shall discuss the details in the chapter devoted to the multilayer neural networks.

1.4.7 Learning CMAC

The Cerebellar Model Articulation Controller (CMAC) [19], [20] can approximate complex non-linear functions much faster than networks using sigmoidal functions. CMAC uses the concept of locally tuned overlapping receptive fields. During the learning process only a small subset of the network's memory is adjusted at each training point.

CMAC is intended for effective storage, restoration, and interpolation of multidimensional functions. The operating principles of the CMAC can be briefly stated as follows:

An input vector s containing the values of the arguments of the function to be stored is coded by a special algorithm so that the input vector s is associated with a fixed number ρ of active associative memory cells in the CMAC, where the total number of cells is p and $p \gg \rho$. The relationship between the input vector s and the active memory cells is uniquely described by a p-dimensional association vector u. The components u_i, $i=1,...,p$, may take the values zero or one. Zeros are located in those positions in the vector u corresponding to inactive associative cells, and ones are placed in positions corresponding to active associative cells. We will call the one and zero-elements of the association vector active and inactive elements, respectively. The associative memory cell with index i contains a number w_i called a weight. The weights w_i form a p-dimensional weight vector w. The elements of the association vector u and the weight vector w are ordered in the same way according to the indices of CMAC memory cells. The weights w_i are chosen so that the average value of the active weights is equal to the value of the function to be stored. This method of organising the memory in CMAC (in which information on the values of a stored function is stored in the form of cells of associative memory) provides the important capability of interpolation for the values of the stored function, i.e., the possibility of automatically predicting a value of the function that has not yet been stored in the CMAC.

We must distinguish two different processes in the CMAC: the training process, in which an algorithm uses changes in the arguments of a function and its values to determine the weights w_i , and the restoration process, in which an input vector s is used to restore or estimate the values of the function y^*. During the training process, when an CMAC at the n-th step is presented with input vector $u(n)$ and a corresponding function value $y^*(n)$, the weight vector w is corrected so that it takes the value $w(n)$ in the n-th step. The training process is constructed so that the sequence $w(n)$ approaches a stationary value as the number of observations n increases.

The algorithm for training the CMAC was proposed by Albus in [20]. This algorithm is recursive, so the correction to $w(n-1)$ computed for the weight vector based on $n-1$ measurements is computed according to the following rule: when the n-th value of the input vector $s(n)$ and the corresponding value of the function $y^*(n)$ are received a coding algorithm is used to compute the p-dimensional association vector $u(k)$ from the input vector $s(k)$. Consequently, active and inactive associative elements of the vector $x(k)$ are determined. The arithmetic average of the active weights of the vector $w(k-1)$ is subtracted from the value of the function $y^*(n)$ stored and the difference obtained (correction) is added to the values of the active weights. The inactive values of the weight vector are not changed. Albus' training algorithm can be written analytically in the form

$$w_i(n) = w_i(n-1) + \left(y^*(n) - \sum_j w_j(n-1)/\rho \right),$$

$$\left(\forall\, i,\, u_i(k) = 1;\ \forall\, j,\, u_j(k) = 1 \right),$$

(1.31)

$$w_i(n) = w_i(n-1),\ \left(\forall\, i,\, u_i(k) = 0 \right).$$

(1.32)

The sequence $w(n)$ in (1.31)-(1.32) is constructed so that the arithmetic average of the corrected weights is equal to the stored value of the function $y^*(n)$ in the n-th step.

References

1. DeGroot M. H.," Optimal statistical decisions", New York, McGraw-Hill Comp., 1970.
2. Censor Y. ," Row-action methods for huge and sparse systems and their application", SIAM REVIEW, vol. 23, No. 4, pp. 444-466, October 1981.
3. Polyak B. T., "Introduction to optimisation", Optimisation Software, Inc., Publications Division, New York, 1987.
4. Tsypkin Ya. Z. , "Adaptation and learning in automatic systems", New York and London: Academic Press, 1971.
5. Tsypkin Ya. Z. , "Foundation of theory of learning systems", New York and London: Academic Press, 1973.
6. Widrow B., Lehr M. ,"30 Years of Adaptive Neural Networks: Perceptron, Madaline, and Backpropagation", Proceedings of the IEEE, Vol. 78, No. 9, 1990, pp. 1415-1440
7. Aved'yan E. D., Simsar'yan R. A., "Adaptive functional transformer in the problem of determining the parameters of an engineering process from indirect indications", Automation and remote control, No. 11. pp. 1797-1801, 1969
8. Tsypkin Y. Z., Aved'yan, "Optimal off-line signal processing", Computers Electrical Engineering, vol. 19, No. 1, pp. 41-46, 1993
9. Aved'yan E. D.," Adaptive filter using time-delay elements", Automation and remote control, No. 9, pp. 1438-1441, 1969
10. Hunt K. J. , Sbarbaro D., Zbikowski R. and Gawthrop, " Neural networks for control systems - a survey", Automatica, vol. 28, No. 6, pp. 1083-1112, 1992
11. Pao Y. H. ,"Adaptive pattern recognition and neural networks", Addison Wesley Publishing Company, Inc., 1989.
12. Chen S., Billings S. A., Grant P. M., "Non-linear system identification using neural networks", Int. J. Control, vol. 51, No. 6, pp. 1191-12141, 1990.
13. "Neural networks for control", Ed. by Miller W. T., Sutton R. S., Werbos P. J., The MIT Press, 1990.
14. "Neural networks for control and systems", Ed. by Warwick K., Irwin G. W., Hunt K. J., Peter Peregrinus Ltd., 1988.
15. Horne B., Jamshidi M., Vadice N., "Neural networks in robotics: a survey", J. of Intelligent and Robotic Systems, No. 3, pp. 51-66, 1990.
16. Rumelhart D. E., Hinton G. E., Williams R. J., "Learning internal representations by error propagation", Parallel Distributed Processing, V.1, ch.8, D. E. Rumelhart and J. L. McClelland, Eds., Cambridge, MA: MIT Press, 1986.
17. Werbos P. J., "Backpropagation through time: what it does and how to do it", Proc. of the IEEE 1990, vol. 78, No. 10, pp. 1550-1560.
18. Narendra K., Parthasarathy K., "Identification and control of dynamic systems using neural networks", IEEE Tr. on NN, 1990, No. 1, pp. 4-26.

19. Albus, J. S.," A new approach to manipulator control: The cerebellar model articulation controller (CMAC)". Trans. the ASME, J. of Dynamic Systems, Measurement, and Control, 97, pp. 220-227, 1975.
20. Albus, J. S.," Data storage in the cerebellar model articulation controller (CMAC)". Trans. of the ASME, J. of Dynamic Systems, Measurement, and Control, 97, pp. 228-233, 1975.

2 Deterministic Algorithms

2.1 Simple Projection Algorithms in Spaces With Different Norms (Structure, Convergence, Properties)

In the first chapter we have considered some examples of learning systems. It follows from these examples that any learning system is trying to imitate the teacher's behaviour. This can achieved by means of construction of a model of the teacher's instructions. The more we know about the teacher's behaviour the better the possible quality of the learning system.

In this chapter it is supposed that the teacher's behaviour is described by the following equation

$$y^*(n) = \sum_{i=1}^{N} w_i^* u_i(n) = w^{*T} u(n),$$
(2.1)

where vector w^* is an unknown vector of parameters, whereas the input vector $u(n)$ and the teacher's instruction $y^*(n)$ are known to the learning system.

It is possible to choose the model of the teacher's instruction for the learning system in different ways. The best model in this case is the linear combiner (chapter one), described by the equation

$$y(n) = \sum_{i=1}^{N} w_i u_i(n) = w^T u(n),$$
(2.2)

because its structure coincides with that of the teacher's instruction. Clearly, we are facing a typical identification problem, namely the problem of estimating vector w^*.

In this simple case, we can express the policy of the learning system as a task of solving a system of linear equations, rather then minimising a performance index. More precisely, we have to solve the system of linear equations

$$y^*(m) = \sum_{i=1}^{N} w_i u_i (m) = u^T(m)w, \; m = \overline{1,n} . \tag{2.3}$$

with respect to unknown vector w^*.

The solution $\overline{w}$ of the system (2.3), which doesn't necessarily coincide with w^* (think of the case when system (2.3) is underdetermined) is the optimal value of the learning system memory.

It is possible to solve system (2.3) using matrix inversion or pseudo inversion. In the case when system (2.3) is huge and sparse this direct approach is very inconvenient. This is a typical situation for the learning system CMAC where the dimension of vector u can be of order 10^5. Moreover it is possible that vector w^* is slowly time-varying, so that it is also impossible to get a solution by matrix inversion.

Fortunately, in the field of linear problems, linearly constrained optimisation problems and interval convex programming problems there are some methods, namely the row-action methods, see [1], which are very convenient for our task. The main feature of row-action methods is that they are iterative procedures and that they use only one row of the matrix at each step of iteration. The first to propose a row-action algorithm was the Polish mathematician Stefan Kaczmarz [2] in 1937. We can find an English version of Kaczmarz's article and an interpretation of Kaczmarz algorithm in connection with learning control systems in [3],[4]. The Kaczmarz algorithm has been rediscovered more than once, for example by Albus, 1975.

2.1.1 Construction Of Kaczmarz Algorithm

Introduction: The i-th equation of system (2.3) can be seen as a hyperplane $H_i = \left\{ w \in R^N, \; y^*(i) = u^T(i) \, w \right\}$ in the N-dimensional Euclidean space. The point $w(0)$, the initial approximation of the solution of the system, is arbitrarily assigned in this space. This point is orthogonally projected onto the first hyperplane H_1 and this projection is equal to $w(1)$. Then the projection $w(1)$ itself is projected onto the second hyperplane H_2, etc. The projection procedure describes an iterative process towards the solution of the system of equation (2.3), and it corresponds to the Kaczmarz algorithm in which the estimates at the n-th and $(n\text{-}1)$-th steps of iterations are related as follows:

$$w(n) = w(n-1) + \frac{y^*(n) - u^T(n)w(n-1)}{u^T(n)u(n)} u(n). \qquad (2.4)$$

Derivation of the Kaczmarz Algorithm: Let us reformulate the construction of this algorithm as follows: If we obtain at the $(n\text{-}1)$-th step an estimate $w(n\text{-}1)$ in the parameter space W, then following estimate $w(n)$, after receiving the n-th equation of the system $(2.3)(\ y^*(n), u(n)\ )$, will be obtained by minimisation of the Euclidean distance $\rho_2(w(n), w(n-1))$ between these estimates under the condition that the estimate $w(n)$ satisfies the n-th equation of system (2.3). This can be expressed analytically as

$$\min_{w(n)} \rho_2(w(n), w(n-1)) = \min_{w(n)}((w(n) - w(n-1))^T(w(n) - w(n-1)))^{\frac{1}{2}} \qquad (2.5)$$

under the condition

$$y^*(n) - u^T(n)\ w(n) = 0. \qquad (2.6)$$

It easy to show that algorithm (2.4) yields the solution of problem (2.5)-(2.6). In fact, problem (2.5)-(2.6) is a linearly constrained problem which can be easily solved by introducing the Lagrange function :

$$L(w(n), \lambda(n)) =$$
$$\rho_2(w(n), w(n-1)) + \lambda(n)(y^*(n) - u^T(n)\ w(n)). \qquad (2.7)$$

In the Lagrange function (2.7) $\lambda(n)$ is a Lagrange multiplier. Obviously, the solution of problem (2.5)-(2.6) satisfies the following system of equations for $w(n)$ and $\lambda(n)$:

$$\nabla_{w(n)}L(w(n), \lambda(n)) = \rho_2^{-1}(w(n), w(n-1))(w(n) - w(n-1)) - \lambda(n)u(n) = 0 \qquad (2.8)$$

$$\nabla_{\lambda(n)}L(w(n), \lambda(n)) = y^*(n) - u^T(n)\ w(n) = 0. \qquad (2.9)$$

By multiplying both sides of Eq.(2.8) by $u^T(n)$ from the left and using condition (2.9) we obtain $\lambda(n)$:

$$\lambda\ (n) = \frac{y^*(n) - u^T(n)\ w\ (n-1)}{u^T(n)\ u\ (n)} \rho_2^{-1}(w(n),\ w\ (n-1)).$$ (2.10)

Substitution of (2.10) into (2.8) yields Kaczmarz algorithm (2.4) (Q-algorithm).

Monotonic Property of the Kaczmarz Algorithm: The main property of Kaczmarz's algorithm is that the estimate w(n) at the step n is not worse than the estimate w(n-1) at the step n -1.

In fact, let us subtract vector w^* from both sides of Eq.(2.4) , where w^* is the solution of the linear system of equations (2.3). Taking into account the fact that $y^*(n) = u^T(n)\ w^*$, we get

$$w\ (n) -\ w^* = \left(I - \frac{u(n)u^T(n)}{u^T(n)u(n)}\right)(w(n-1) -\ w^*)$$

or

$$\Delta\ w(n) = \left(I - \frac{u(n)u^T(n)}{u^T(n)u(n)}\right)\Delta\ w(n-1).$$ (2.11)

In Eq.(2.11) I is NxN unit matrix and $\Delta\ w(n) = w\ (n) -\ w^*$. If we multiply both sides of Eq.(2.11) by $\Delta^T w(n)$ from the left, we obtain

$$\Delta^T w(n)\Delta\ w(n) = \Delta^T w(n-1)\left(I - \frac{u(n)u^T(n)}{u^T(n)u(n)}\right)^2 \Delta\ w(n-1).$$ (2.12)

giving:

$$L^2(n) = L^2(n-1) - \frac{(u^T(n)\Delta\ w(n-1))^2}{u^T(n)u(n)} \le L^2(n-1) ,$$ (2.13)

where $L^2(n)$ is the squared distance between point w(n) and the solution w^* and $L^2(n-1)$ is the squared distance between point w(n-1) and the solution w^*. It follows from (2.13) that the sequence $L^2(n)$ is monotonically decreasing and bounded.

Projection Algorithms In Non-Euclidean Spaces: It is possible to extend the approach described above to non-Euclidean spaces [5]. Let us consider a metric space, in which the distance between two vectors is given by the expression

$$\rho_q(w(n), w(n-1)) = \left(\sum_{i=1}^{N} |w_i(n) - w_i(n-1)|^q \right)^{1/q} \tag{2.14}$$

where $1 \leq q < \infty$. In this case, the problem (2.5)-(2.6) can be written as follows:

$$\min_{w(n)} \rho_q(w(n), w(n-1)) = \min_{w(n)} \left(\sum_{i=1}^{N} |w_i(n) - w_i(n-1)|^q \right)^{1/q} \tag{2.15}$$

under the condition

$$y^*(n) - u^T(n) \, w(n) = 0 . \tag{2.16}$$

Especially interesting situations arise when $q \to \infty$ and $q = 1$. In the first case we have the supremum, or l_∞, norm

$$\|w\|_\infty = \max_i |w_j| . \tag{2.17}$$

In the second case we have the l_1 norm, or absolute distance

$$\|w\|_1 = \sum_{i=1}^{N} |w^*| . \tag{2.18}$$

Consider now problem (2.15)-(2.16) with $q = 2k$ and $q = \dfrac{2k}{2k-1}$ and take the limit for $k \to \infty$ for the cases (2.17) and (2.18) respectively. Then we obtain the C-algorithm for the *supremum norm* (case (2.17))

$$w(n) = w(n-1) + \frac{y^*(n) - u^T(n)w(n-1)}{u^T(n)sign(u(n))} sign(u(n)), \tag{2.19}$$

where

$$sign(u(n)) = (sign(u_1(n)), ..., sign(u_N(n)))^T \qquad (2.20)$$

and

$$sign(u_i(n)) = \begin{cases} 1 & if \quad u_i(n) > 0 \\ 0 & if \quad u_i(n) = 0 \\ -1 & if \quad u_i(n) < 0 \end{cases}. \qquad (2.21)$$

It follows from (2.21) that $u^T(n)sign(u(n)) = \sum_{i=1}^{N} |u_i(n)|$.

For the l_1 norm (case (2.18)) we obtain the O-algorithm:

$$w_j(n) = w_j(n-1) + \frac{y^*(n) - u^T(n)w(n-1)}{u_j(n)},$$

where $|u_j(n)| = \max_i |u_i(n)|$, $\qquad (2.22)$

$w_i(n) = w_i(n-1)$ for all $i \neq j$.

A comparison of the all three algorithms (Q, C and O-algorithms) shows that, from a computation point of view the O-algorithm is the simplest one because only one component of vector w is changed at each iteration step. This component corresponds to the component of vector $u(n)$ with maximum absolute value.

The C-algorithm was proposed by Naguma and Noda [6]. For all three algorithms a nice geometrical interpretation can be derived.

Projection Algorithms With Partial Relaxation: The Q, C and O-algorithms are full relaxation algorithms, i.e. at each iteration step of the iterations the point $w(n)$ is the solution of the n-th equation of system (2.3): $y^*(n) \equiv u^T(n) w(n)$ (i.e., the point $w(n)$ is lying on the hyperplane H_n). Let us introduce partial relaxation algorithms. This can be easily done by multiplying the correction term of the Q, C and O-algorithms by some constant γ which must lie in the interval $0 < \gamma < 2$. The properties of the algorithms change by introducing the multiplier γ, but the convergence property of the Q-algorithm will remain unchanged: the distance between point $w(n)$ and the solution cannot increase. The Q, C and O-algorithms with partial relaxation are as follows:

$$w(n) = w(n-1) + \gamma \, \frac{y^*(n) - u^T(n)w(n-1)}{u^T(n)u(n)} u(n), \qquad (2.23)$$

$$w(n) = w(n-1) + \gamma \, \frac{y^*(n) - u^T(n)w(n-1)}{u^T(n)sign(u(n))} sign(u(n)), \qquad (2.24)$$

$$w_j(n) = w_j(n-1) + \gamma \, \frac{y^*(n) - u^T(n)w(n-1)}{u_j(n)}, \quad \text{where} \quad |u_j(n)| = \max_i |u_i(n)|$$

and $w_i(n) = w_i(n-1)$ for all $i \neq j$, $0 < \gamma < 2$. $\qquad (2.25)$

Parameter γ will play a very important role in cases when the learning system doesn't coincide with the teacher's behaviour (teacher's instructions are corrupted by noise or are time-varying).

2.1.2 Convergence

Convergence properties of these algorithms can be stated in various ways, statistical or otherwise. Hereafter, we shall make use of the concept of convergence in the statistical mean-squares sense, i.e. we will say that the process $w(n)$ converges to w^* if it satisfies the condition $\lim_{n \to \infty} E\{(w(n) - w^*)^T (w(n) - w^*)\} = 0$, where E denotes the statistical expectation operator.

The convergence depends greatly on the properties of the input process $u(n)$: for instance, the Q-algorithm will not converge if the process $u(n)$ quickly settles to a constant value or if it varies very slowly. The situation with the C and O-algorithms is more obvious. The C-algorithm will not converge in any sense if all the components of the vector $u(n)$ are positive (or negative) for all n. Hence, we must carry out a priori investigations before applying these learning algorithms to a given learning system.

In the particular case when the input process is a white noise process (the components of the vector $u(n)$ are mutually uncorrelated stationary random processes ($E\{u_i(m)u_j(n)\} = 0$; $i \neq j$, $m \neq n$), with equal variances $\sigma^2 = E\{u_i(n)\}$), it is relatively easy to obtain various characteristics of the algorithms (2.23)-(2.25). We shall here give the characteristics of the Q-algorithm.

The Convergence Time: In order to find the convergence conditions for the Q-algorithm let us subtract w^* from both sides of the Eq.(2.23). Using the fact that

$y^*(n) \equiv u^T(n) w(n)$ we   can   rewrite   Eq.(2.23)   in   the   variable
$\Delta w(n) = w(n) - w^*$:

$$\Delta w(n) = \left(I - \gamma \frac{u(n)u^T(n)}{u^T(n)u(n)} \right) \Delta w(n-1) . \tag{2.26}$$

Multiplying Eq.(2.26) by $\Delta^T w(n)$ from the left we get

$$\Delta^T w(n)\Delta w(n) =$$

$$= \Delta^T w(n-1) \left(I - (2\gamma - \gamma^2) \frac{u(n)u^T(n)}{u^T(n)u(n)} \right) \Delta w(n-1) . \tag{2.27}$$

Applying the statistical expectation operator to the left and right sides of the Eq.(2.27), and remembering that process $u(n)$ is a white noise and the statistical expectation $E\left\{ \dfrac{u(n)u^T(n)}{u^T(n)u(n)} \right\} = N^{-1}I$ we can write

$$l^2(n) = (1 - (2\gamma - \gamma^2)N^{-1})l^2(n-1) , \tag{2.28}$$

where $l^2(n) = E\left\{ \Delta^T w(n)\Delta w(n) \right\}$. Now we have an equation that describes the convergence property of algorithm (2.23). The process (2.23) will converge if $l^2(n) \to 0$ when $n \to \infty$. This is possible if

$$0 < (1 - (2\gamma - \gamma^2)N^{-1}) < 1 . \tag{2.29}$$

The condition (2.29) is fulfilled when $0 < \gamma < 2$ for any N. The rate of convergence is given by the parameter $q = (1 - (2\gamma - \gamma^2)N^{-1})$. The smaller the parameter q, the greater the convergence rate. It follows immediately from the solution of equation (2.28) that

$$l^2(n) = q^n l^2(0) = e^{n \ln q} l^2(0) = e^{-\frac{n}{-1/\ln q}} l^2(0) = e^{-\frac{n}{T_\varrho}} l^2(0) , \tag{2.30}$$

where $T_\varrho = -1/\ln q$.

In Eq.(2.30) $l^2(0)$ is the deviation of the initial value $w(0)$ from the solution $w^*(n)$: $l^2(0) = E\{\Delta^T w(0)\Delta w(0)\}$. We can estimate the convergence time of the process $w(n)$ as $t_{trans} = (3 \text{ to } 4) T_Q$.

The Influence of Noise: Assume that the teacher's instructions are corrupted with a white noise signal $\xi(n)$, with variance σ_ξ^2, and that the teacher's instruction and the noise are mutually uncorrelated. Then, the teacher's instructions have the form:

$$y^*(n) = w^{*T} u(n) + \xi(n).$$

It is easy to show that the estimates $w(n)$ are asymptotically unbiased for $n \to \infty$:

$$E(w(n) - w^*) = 0. \tag{2.31}$$

The behaviour of the variance $l^2(n)$ of the estimate $w(n)$ can be described by equation

$$l^2(n) = (1 - (2\gamma - \gamma^2)N^{-1})l^2(n-1) + \gamma^2 \sigma_\xi^2 \mu_1, \tag{2.32}$$

where $\mu_1 = E\left(\dfrac{1}{u^T(n)u(n)}\right)$. It follows from Eq.(2.31) that variance $l^2(n)$ asymptotically tends to $\sigma_w^2 = \sigma_\xi^2 \mu_1 \dfrac{\gamma}{2-\gamma}$. This means that, after a transient period, the estimate $w(n)$ will be situated in a zone centred on the point w^*.

Non-Stationary Teacher's Instructions (Tracking Regime): We will consider here the case when the teacher's instructions are non stationary, i.e. they are still described by equation (2.1) but the solution w^* is a time-dependent function $w^* = w^*(n)$. We employ the usual algorithms for the estimation of the w^*. What will happen in this case?

Let us suppose, for example, that function $w^*(n)$ is linear

$$w^*(n) = a n + b, \tag{2.33}$$

where $a, b \in R^N$ are unknown vectors.

It can be shown that after the transient period the statistical expectation of the estimate $w(n)$ is $E\{w(n)\} = a n + b - \Delta a$, where $\Delta a = (\gamma^{-1} N - 1)a$ is a bias, proportional to vector a. Hence the Q-algorithm has tracking properties. The C and O-algorithms have the same properties. It is possible to show that tracking properties for the C and O-algorithms are worse than for the Q-algorithm.

The characteristics of other algorithms can be derived using the same approach employed for the Q-algorithm.

2.2 Modified Projection Algorithms With a High Rate of Convergence

2.2.1 Construction

For the construction of algorithms with a higher rate of convergence, let us extend the above technique by requiring that the estimate $w(n)$ should satisfy not only the n-th equation, but also the last k equations of the system of equations (2.3) under condition (2.5). This approach [7] means that the point $w(n-1)$ is projected orthogonally onto the intersection of the last k hyperplanes of system (2.3). It is evident that if the vectors $u(n-m)$, $m = \overline{0, k-1}$ are linearly independent , number k must be not greater than the dimension N of vector w.

Thus, the criterion corresponding to the modified Kaczmarz algorithm is

$$\min_{w(n)} \rho_2 (w(n), w(n-1)) = \min_{w(n)} ((w(n) - w(n-1))^T (w(n) - w(n-1)))^{\frac{1}{2}} \quad (2.34)$$

under the condition

$$y^* (n-m) - u^T (n-m) w(n) = 0, \; m = \overline{0, k-1} . \quad (2.35)$$

We can again introduce the Lagrange function for problem (2.34)-(2.35):

$$L(w(n), \lambda (n)) =$$

$$\rho_2 (w(n), w(n-1)) + \sum_{m=0}^{k-1} \lambda_m (n)(y^* (n-m) - u^T (n-m) w(n)) \quad (2.36)$$

In Eq.(2.36) the $\lambda_m(n)$, $m = \overline{0, k-1}$ are Lagrange multipliers. The minimisation of the Lagrange function (2.36) yields the following system of equations

$$w(n) - w(n-1) - \sum_{m=0}^{k-1} \lambda'_m(n)u(n-m) = 0, \tag{2.37}$$

$$y^*(n-m) - u^T(n-m)\,w(n) = 0, \quad m = \overline{0, k-1},$$

where $\lambda'_m(n) = \rho_2(w(n), w(n-1))\lambda_m(n)$.

Let us introduce the following notation:

$U_k(n) = (u(n), u(n-1), ..., u(n-k+1))$ is a matrix of dimension $N \times k$; $Y_k^{*T}(n) = (y^*(n), y^*(n-1), ..., y^*(n-k+1))$ and
$\Lambda_k^T(n) = (\lambda'_0(n), \lambda'_1(n), ..., \lambda'_{k-1}(n))$ are k-dimensional row vectors.
With the use of this notation the system of equations (2.37) takes the form

$$w(n) - w(n-1) = U_k(n)\Lambda(n), \tag{2.38}$$

$$Y^*(n) - U_k^T(n)\,w(n) = 0. \tag{2.39}$$

To find the unknown vectors $w(n)$ and $\Lambda(n)$, let us multiply both sides of Eq.(2.38) by $U_k^T(n)$ from the left:

$$U_k^T(n)\ w(n) - U_k^T(n)\ w(n-1) = U_k^T(n)\ U_k(n)\Lambda(n). \tag{2.40}$$

By virtue of condition (2.35), which holds for any $n > k$ and hence also for $n-1$, the vector

$$(U_k^T(n)\ w(n-1))^T =$$
$$= (u^T(n)w(n-1), u^T(n-1)w(n-1), ..., u^T(n-k+1)w(n-1))$$

is equal to

$$(u^T(n)w(n-1), \ y^*(n-1), \ y^*(n-2), ..., \ y^*(n-k+1)).$$

Hence, the expression in the left-hand side of Eq.(2.40) is equal to

$(y^*(n) - u^T(n) \, w(n-1)) e^T$,

where $e^T = (1, 0, 0, ..., 0)$ is the k-dimensional unit row vector, since $(U_k^T(n) \; w(n))^T = (y^*(n), y^*(n-1), ..., y^*(n-k+1))$ by virtue of Eq.(2.35). Thus Eq. 40 takes the form

$(y^*(n) - u^T(n) \, w(n-1)) e = U_k^T(n) \, U_k(n) \Lambda(n)$,

and we obtain an expression for the unknown vector of Lagrange multipliers

$$\Lambda(n) = (U_k^T(n) \, U_k(n))^{-1} e \, (y^*(n) - u^T(n) \, w(n-1)) \qquad (2.41)$$

under the condition that the matrix $U_k^T(n) U_k(n)$ is not singular. By substituting (2.41) into (2.38), we obtain the unknown vector $w(n)$.

$$w(n) = w(n-1) + U_k(n)(U_k^T(n) \, U_k(n))^{-1} e \, (y^*(n) - u^T(n) \, w(n-1)). \quad (2.42)$$

The computation of $w(n)$ is a modified Kaczmarz algorithm which makes it possible to calculate a new estimate that satisfies the condition (2.34) and (2.35), on the basis of the previously obtained estimate $w(n-1)$. The algorithm has been obtained in this form in [8]. For small values of k it is also possible to invert the matrix appearing in (2.42). Thus for $k=1$, $U_1(n) \, (U_1^T(n) \, U_1(n))^{-1} e = u(n)/u^T(n)u(n)$, and the algorithm evidently coincides with the Kaczmarz algorithm. For $k=2$,

$$U_2(n)\left(U_2^T(n)U_2(n)\right)^{-1} e =$$
$$\frac{u(n)\left(u^T(n-1)u(n-1)\right) - u(n-1)\left(u^T(n)u(n-1)\right)}{\left(u^T(n)u(n)\right)\left(u^T(n-1)u(n-1)\right) - \left(u^T(n)u(n-1)\right)^2}$$

For larger values of k a direct calculation of the estimates $w(n)$ by algorithm (2.42) is costly in terms of computational load.

In order to give a more convenient form to algorithm (2.42) for calculation purposes, let us decompose the matrix $U_k(n)$ occurring in the algorithm (2.42) into blocks: $U_k(n) = (u(n) | U_{k-1}(n-1))$. By virtue of the rules of multiplication and inversion of block matrices, the algorithm (2.42) takes the following form

$$w(n) = w(n-1) + \frac{R_{k-1}(n-1)u(n)}{u^T(n)R_{k-1}(n-1)u(n)} \; (y^*(n) - u^T(n) \, w(n-1)) \qquad (2.43)$$

where

$$R_{k-1}(n-1) = I - U_{k-1}(n-1)(U_{k-1}^T(n-1)U_{k-1}(n-1))^{-1}U_{k-1}^T(n-1) \qquad (2.44)$$

is an $N \times N$- matrix and I is the unit matrix.

The matrix $R_{k-1}(n-1)$ in Eq.(2.43) and (2.44) can be computed recursively. In fact, simple computations show

$$R_i(j) = R_{i-1}(j-1) - \frac{R_{i-1}(j-1)u(j)u^T(j)R_{i-1}(j-1)}{u^T(j)R_{i-1}(j-1)u(j)}, \qquad (2.45)$$

where $i = \overline{1, k-1}$; $j = n - k + i$ and $R_0(.) = I$ is a unit matrix.

2.2.2 Transient Mode

Calculation of the estimates in accordance with Eqs.(2.43) and (2.45) is possible only after instant $n \geq k$, by virtue of the condition (2.35). Up to this instant, the estimates can be calculated with an algorithm constructed in accordance with criterion (2.34), (2.35) in which, however, parameter k is variable, i.e. $k=n$. In this case we perform an orthogonal projection onto the first hyperplane, then onto the intersection of the first and the second hyperplane, then onto the intersection of the first, the second, and the third hyperplane, etc., up to the orthogonal projection onto the intersection of the first k hyperplanes. The corresponding computation procedure follows from (2.43) and (2.45) with $k=n=i$.

$$w(n) = w(n-1) + \frac{R(n-1)u(n)}{u^T(n)R(n-1)u(n)} \; (y^*(n) - u^T(n) \, w(n-1)), \qquad (2.46)$$

$$R(n) = R(n-1) - \frac{R(n-1)u(n)u^T(n)R(n-1)}{u^T(n)R(n-1)u(n)}. \qquad (2.47)$$

In algorithm (2.46)-(2.47) $R(0) = I$ and $w(0)$ is arbitrary, $n = 1, 2,..., k$. The algorithm coincides with the first k steps of the recursive least squares algorithm for linearly independent measurements [9].

2.2.3 Properties Of The Estimates

The estimates generated by the algorithm have the property that each new estimate is not further away from the solution than the previous estimate, i.e.,

$$L_k^2(n) = (w(n) - w^*)^T (w(n) - w^*) \le L_k^2(n-1) \ . \tag{2.48}$$

To prove this assertion we shall subtract the vector w^* from the left- and right sides of Eq.(2.46), and multiply the obtained expression from the left by the vector $(w(n) - w^*)^T$. At the end we obtain

$$L_k^2(n) = L_k^2(n-1) - \frac{(u^T(n)(w(n-1) - w^*))^2}{u^T(n) R_{k-1}(n-1) u(n)} \ . \tag{2.49}$$

Since the value

$$\alpha_k(n) = u^T(n) R_{k-1}(n-1) u(n) \tag{2.50}$$

is positive if the k successive vectors $u(n-m)$, $m = \overline{0, k-1}$ are linearly independent (2.48) is proved.

It also follows from (2.49) that the effectiveness of the algorithm increases with k. Indeed, by taking the point $w(n\text{-}1)$ as the estimate at the $(n\text{-}1)$-th step for algorithms with distinct values of k , we obtain $L_k^2(n) \le L_{k-i}^2(n), i = \overline{1, k-1}$, since $\alpha_k(n) \le \alpha_{k-i}(n)$. The last inequality can be obtained using equation (2.45). See also [10].

2.2.4 "Bad" Measurements

In constructing the modified Kaczmarz algorithm, it was implicitly assumed that the k successive vectors $u(n-m)$, $m = \overline{0, k-1}$ are linearly independent. In this case the matrix $U_k^T(n) U_k(n)$ has an inverse , and the estimates are calculated by algorithm (2.46)-(2.47).

Suppose now that the vectors $u(n-m)$, $m = \overline{1, k-1}$ are linearly independent, whereas the vector $u(n)$ is linearly dependent on the previous $k\text{-}1$ vectors $u(n-m)$, $m = \overline{1, k-1}$. Under this assumption the matrix $U_k^T(n) U_k(n)$ is singular and parameter $\alpha_k(n)$ is equal to zero and the estimate $w(n)$ cannot be calculated by equations (2.43)-(2.44).

The linear dependence of vector $u(n)$ on the previous $k\text{-}1$ vectors is equivalent to the fact that the n-th equation of the system (2.3) is a linear combination of the *k-1* previous equations which are satisfied by the estimate $w(n\text{-}1)$ by construction of the algorithm. Hence the estimate $w(n\text{-}1)$ also satisfies the *n*-th equation of the system (2.3). Hence if $\alpha_k(n) = 0$ we shall take $w(n) = w(n\text{-}1)$, and neglect the vector $u(n)$ in subsequent steps of the algorithm.

It is evident from geometrical considerations that the strong linear relationship between the vectors $u(n-m)$, $m = \overline{0, k-1}$ and a small length of these causes an increase in the effect of the noise $\xi(n)$ on the estimate $w(n)$ (remember that $y^*(n) = w^{*T} u(n) + \xi(n)$). These factors manifest themselves in the value of $\alpha_k(n)$, i.e., the smaller the value of $\alpha_k(n)$, the stronger the effect of noise on the estimate. For this reason in the generalised algorithm we shall drop not only the "bad" observation $u(n)$ that cause $\alpha_k(n)$ to vanish, but also observations for which this quantity is smaller than an assigned threshold. The value of the threshold is selected as a compromise, taking into account the largest number of observations $w(n)$ and the magnitude of the variance of the estimates $w(n)$ as a function of the statistical properties of the processes $u(n)$ and $\xi(n)$.

Simulation results for these algorithms can be found in [5] and [7]. They confirm the conclusions that although algorithms with a larger value of *k* have a greater rate of convergence, they are more sensitive to noise than algorithms with a smaller value of *k*.

References

1. Censor Y., "Row-action methods for huge and sparse systems and their applications", SIAM REVIEW, vol. 23. No. 4, pp. 444-466, Oct. 1981.
2. Kaczmarz S., " Angenäherte Auflösung von Systemen linearer Gleichungen", Bulletin International de l'Academie Polonaise des Sciences. Lett A, pp. 355-357, 1937.
3.. Kaczmarz S.," Approximate solution of systems of linear equations", Int. J. Control, vol. 57, No. 6, pp. 1269-1271, 1993.
4. Parks P. C., "S. Kaczmarz", Int. J. Control, vol. 57, No. 6, pp. 1263-1267, 1993.
5. Aved'yan E. D.," Relaxation algorithms for linear plant identification", Ph.D. dissertation, Institute for Control Sciences, Moscow, 1971.
6. Nagumo J. , Noda A., "A learning method for system identification", IEEE Trans. on Automatic Control, vol. AC-12, No. 3, 1967.
7. Aved'yan E. D.," Modified Kaczmarz algorithms for estimating the parameters of linear plants", Automation and Remote Control, No. 5, pp. 674-680, 1978.
8. Aved'yan E. D. " Bestimmung der Parameter linearer Modelle stationärer und instationärer Strecken", Messen, Steuern, Regeln, No. 9, pp. 348-350, 1971.
9. Albert A.," Regression and the Moore-Penrose Pseudoinverse", Academic Press, New York and London, 1972.
10. Gantmacher F. R., " Theory of Matrices", Chelsea, New York, 1960.

Appendix

In this Appendix (Fig.1-Fig.10) we show some figures which illustrate properties of the projection algorithms and some simulation results.

Fig. 2.1 Projections in two-dimensional spaces with Euclidean (A), Supremum (B) and Absolute Distance (C) norms.

Fig. 2.2 Trajectories of the estimates obtained with the Kaczmarz algorithm, the C-algorithm, and O-algorithm in the two-dimensional space

Fig. 2.3-Fig. 2.5 Illustration of full (c) and partial (a), (b) relaxation for the Kaczmarz algorithm, the C-algorithm, and the O-algorithm, respectively.

Fig. 2.6 Simulation results for the Kaczmarz algorithm, $N=5$, with a white noise input signal and a noise-free teacher's instruction.

Fig. 2.7 Simulation results for the Kaczmarz algirithm, $N=5$, with a white noise input signal and a teacher's instruction corrupted by noise.

Fig. 2.8 Simulation results for the Kaczmarz algorithm, N, with a white noise input signal and a teacher's instruction corrupted by noise. Effect of the averaging of the estimates.$=5$

Fig. 2.9 Simulation results for the Kaczmarz algorithm, $N=5$, with a white noise input signal and a noise-free teacher's instruction. Nonstationary case:

$$y^*(n) = \sum_{i=1}^{4} w_i^* u_i(n) + w_5^*(n) u_5(n)$$

Fig. 2.10 Simulation results for the modified Kaczmarz algorithm, $N=10$, with a white noise input signal.

Fig. 2.10(a) Illustration of the effect of parameter k on the rate of convergence, noise-free teacher instruction.

Fig. 2.10(b) and Fig. 2.10(c) Illustration of the effects of different values of thresholds and teacher's instruction corrupted by noise on the estimates

Fig. 2.10(d) Illustration of the tracking possibilities of the modified algorithm with different values of k. The model is described by equation

$$y^*(n) = \sum_{i=1}^{9} w_i^* u_i(n) + w_{10}^*(n) u_{10}(n), \text{ where } w_{10}^*(n) = \sin 2\pi n / 200 .$$

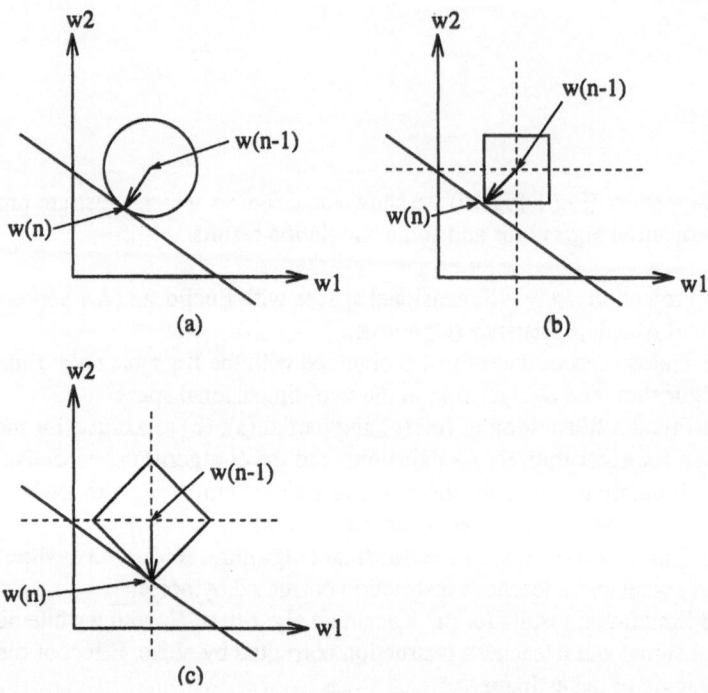

Fig 2.1. Projections in two-dimensional spaces with the Euclidean (a), Supremum (b) and Absolute Distance (c) norms.

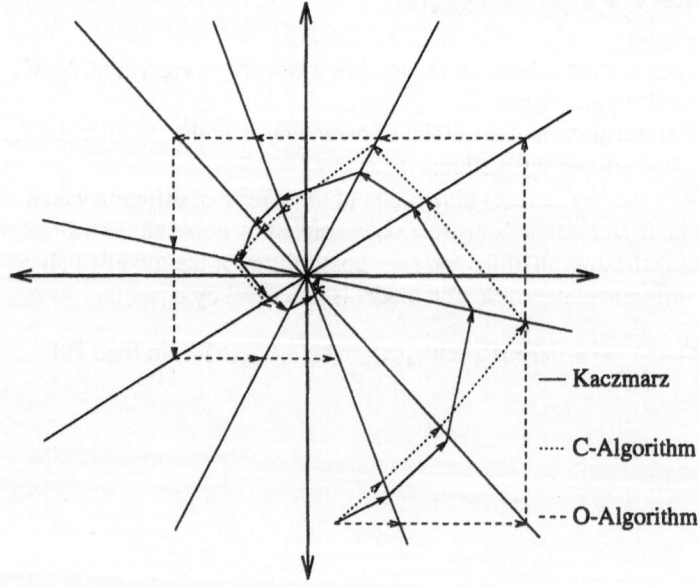

Fig 2.2. Trajectories of estimates obtained with the various norms.

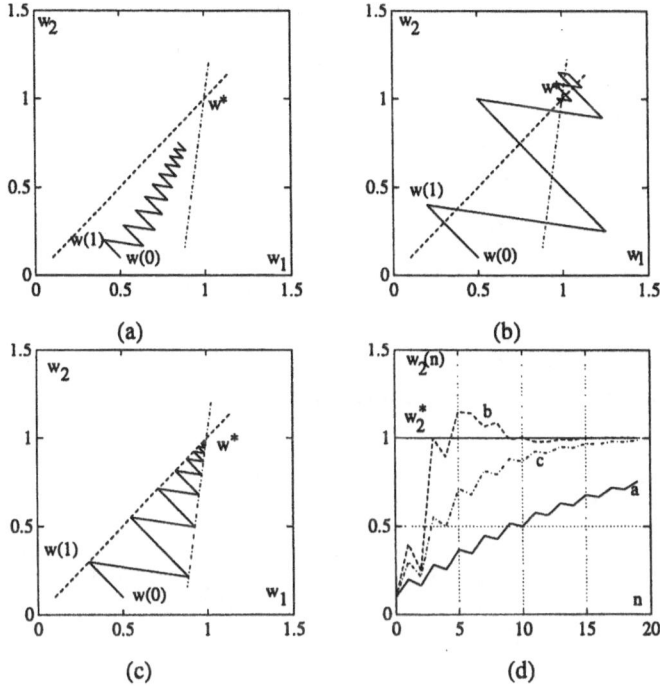

Fig 2.3. Illustration of full relaxation (c) and partial relaxation, $\gamma=0.5$ (a), $\gamma=1.5$ (b) for the Kaczmarz Algorithm.

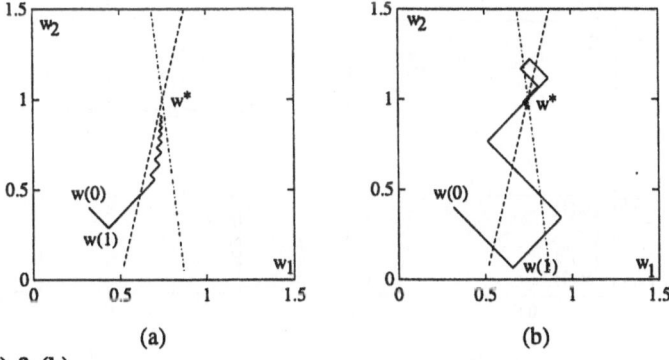

Fig 2.4. (a) & (b)

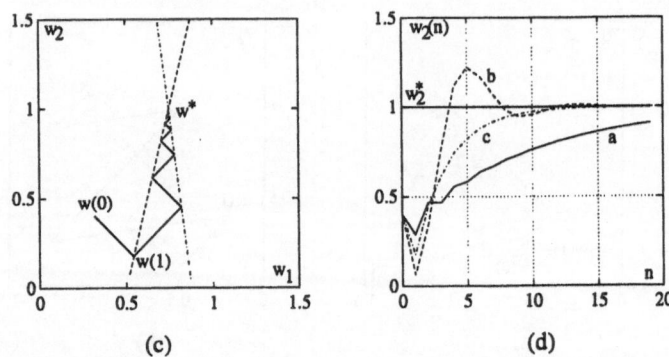

Fig 2.4. Illustration of full relaxation (c) and partial relaxation, γ=0.5 (a), γ=1.5 (b) for the C-Algorithm.

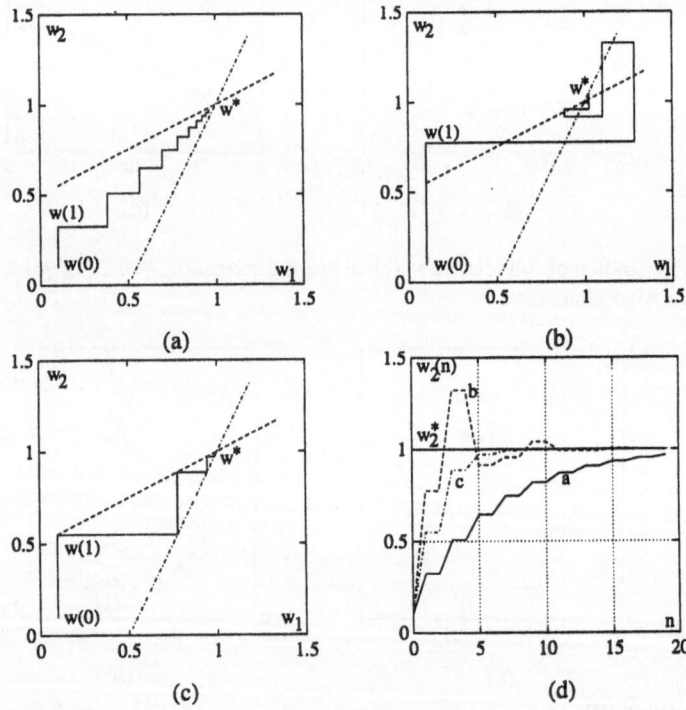

Fig 2.5. Illustration of full relaxation (c) and partial relaxation, γ=0.5 (a), γ=1.5 (b) for the O-Algorithm.

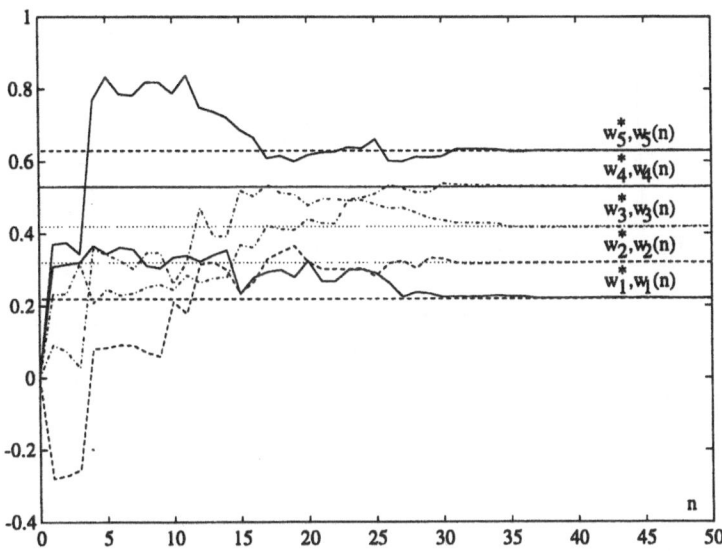

Fig 2.6(a) Simulation results for the Kaczmarz Algorithm

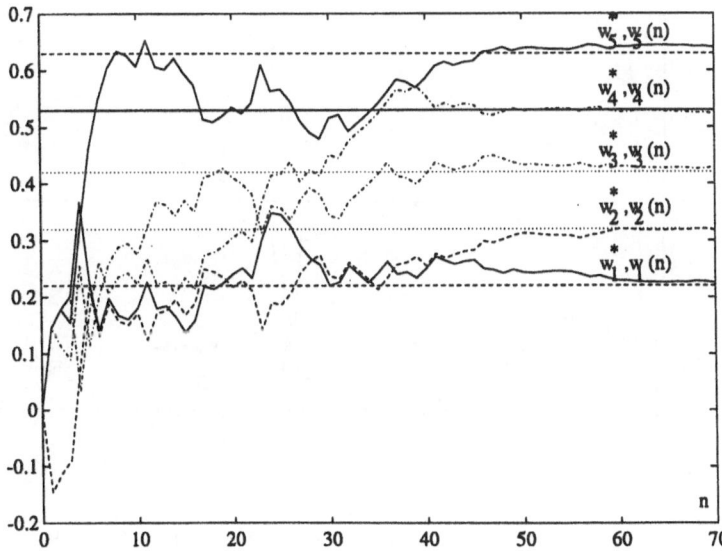

Fig 2.6(b) Simulation results for the Kaczmarz Algorithm

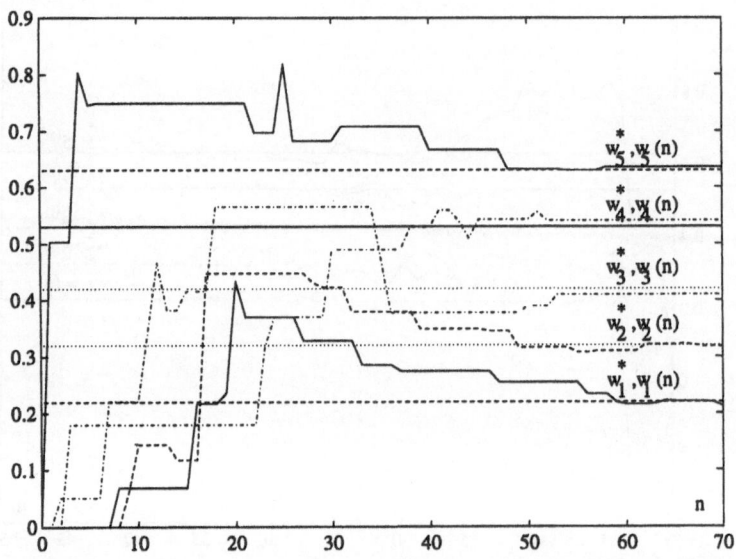

Fig 2.6(c) Simulation results for Kaczmarz Algorithm

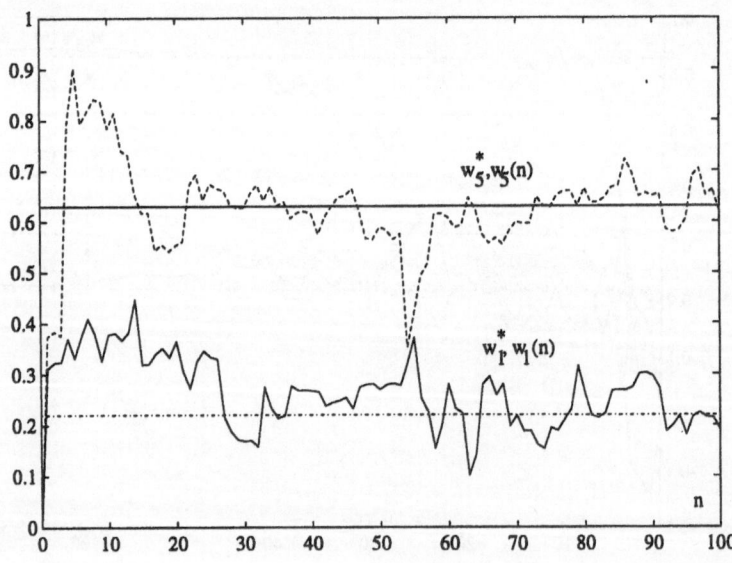

Fig 2.7. Simulation results for Kaczmarz Algorithm with noisy teacher's instructions.

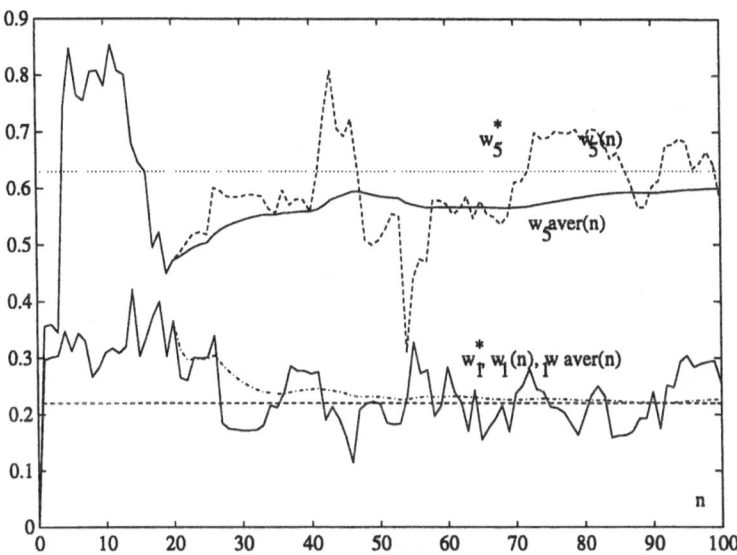

Fig 2.8. Effect of averaging on Kaczmarz Algorithm

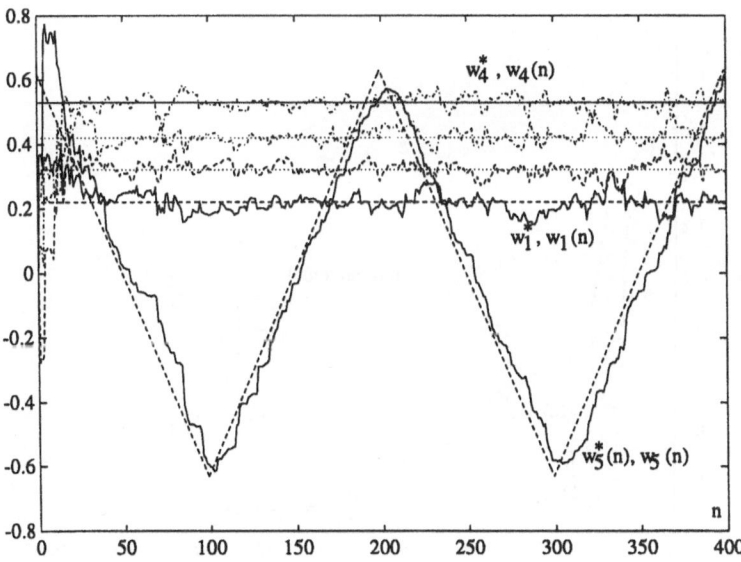

Fig 2.9. Simulation results for Kaczmarz Algorithm. Non-stationary case.

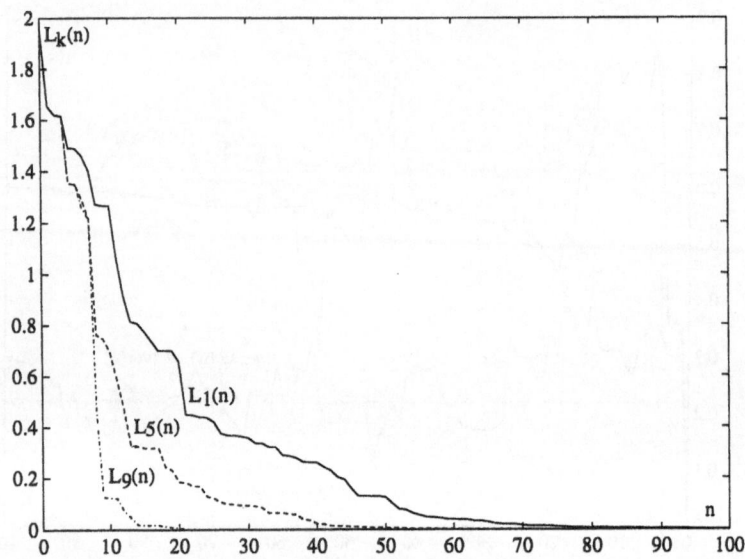

Fig 2.10(a) Simulation of modified Kaczmarz Algorithm

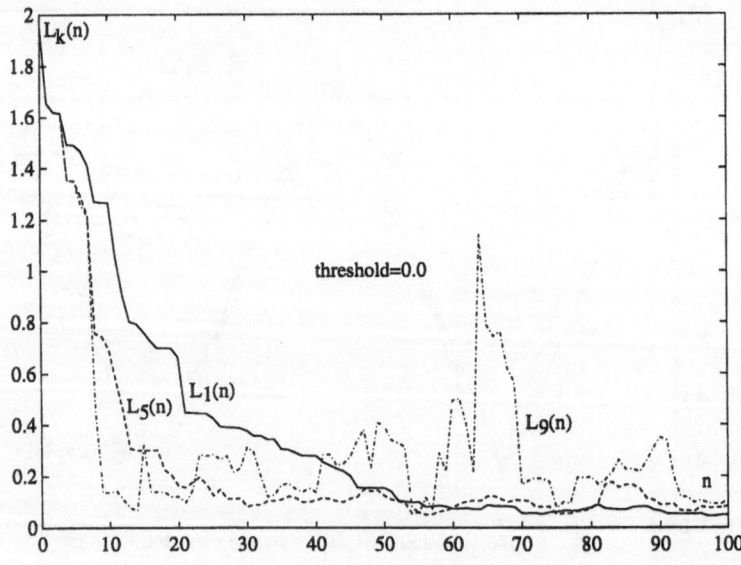

Fig 2.10(b) Simulation of modified Kaczmarz Algorithm.

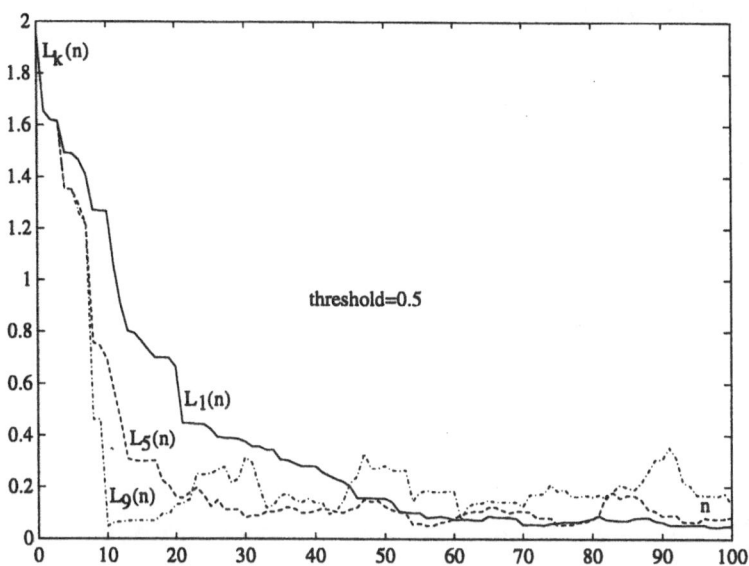

Fig 2.10(c) Simulation of modified Kaczmarz Algorithm

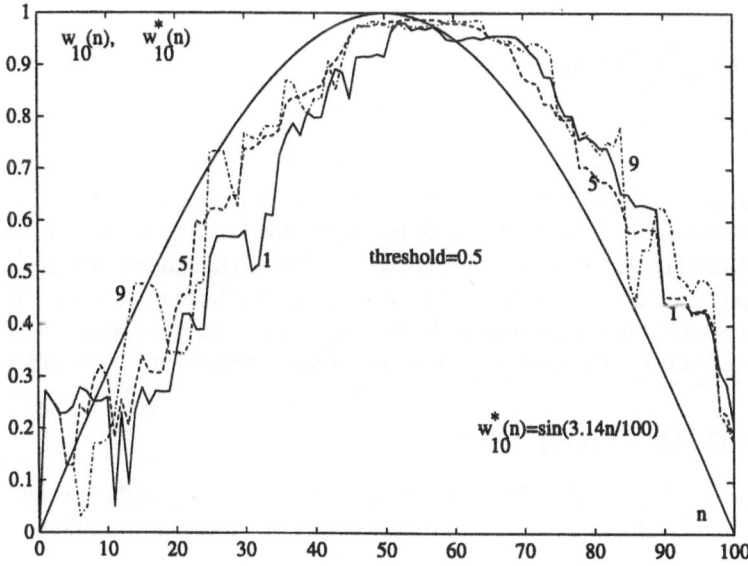

Fig 2.10(d) Simulation of modified Kaczmarz Algorithm

3 Deterministic and Stochastic Algorithms of Optimisation

3.1 Deterministic Methods for Unconstrained Minimisation

In the first part of the chapter we shall assume that the teacher's behaviour is described by a non-linear function and the learning model is also chosen as a non-linear function. It is also supposed that the number P of the input vectors, or patterns, is finite, so that we can use a performance index of the form

$$J(w) = \frac{1}{P} \sum_{m=1}^{P} Q(\varepsilon \ (m;w)). \tag{3.1}$$

The performance index $J(w)$ is not a function of n and we can use various optimisation algorithms to find the optimal value of the memory w^*, e. g. the gradient method, Newton's method, the search method, the Monte-Carlo method and so on [1], depending on what is known of the performance index $J(w)$ and on the presence of constraints on the memory. In conclusion, we have a classical unconstrained or constrained optimisation problem. Here we shall consider two unconstrained minimisation methods: the gradient method and Newton's method.

3.1.1 The Gradient Algorithm

If $J(w)$ is differentiable, has a unique minimum and it is possible to calculate its gradient then its minimum can be obtained with the gradient algorithm:

$$w(n) = w(n-1) - \gamma \ (n) \nabla_w J(w(n-1)), \ n = 1, 2, \dots. \tag{3.2}$$

Here $\gamma(n) \geq 0$ is the step size, $\nabla_w J(w) = \left(\dfrac{\partial J(w)}{\partial w_1}, \ldots, \dfrac{\partial J(w)}{\partial w_N} \right)$ is the gradient of the function $J(w)$ (a column vector, consisting of N partial derivatives $\partial J(w)/\partial w_v, v = \overline{1, N}$ of $J(w)$ with respect to all components of the vector memory w and $w(0)$ is the initial approximation of the solution.

The gradient algorithm attempts to fulfil the necessary extremum condition $\nabla_w J(w) = 0$.

3.1.2 Convergence

Let us consider the gradient method with $\gamma(n) \equiv \gamma = \text{const.}$

Theorem 3.1. Let $J(w)$ be differentiable on R^N and bounded below $J(w) \geq J(w^*) > -\infty$, the gradient $\nabla_w J(w)$ satisfy the Lipschitz condition $\left\| \nabla_w J(u) - \nabla_w J(v) \right\| \leq L \left\| u - v \right\|$ with constant L (where $\|\cdot\|$ is the Euclidean norm in R^N : $\left\| w \right\|^2 = w_1^2 + w_2^2 + \ldots + w_N^2$) and the step size γ satisfy the condition $0 < \gamma < 2/L$, then the function $J(w)$ monotonically decreases: $J(w(n)) \leq J(w(n-1))$ and the gradient tend to zero then:

$$\lim_{n \to \infty} \nabla_w J(w(n)) = 0$$

The proof of the theorem will be not given here. We can easily understand the idea of the theorem in the one dimensional case.

By construction, if the extremum condition does not hold at point $w(n-1)$ ($\nabla_w J(w) \neq 0$), then the value of the function can be decreased by moving to the following point $w(n) = w(n-1) - \alpha \nabla_w J(w)$ for sufficiently small positive α.

The construction of the algorithm can be explained from another point of view. It is well known that the Taylor series of the scalar function $J(w)$ of the N-dimensional argument w in the point $w(n-1)$ has the form:

$$J(w) = J(w(n-1)) + (w - w(n-1))^T \nabla_w J(w(n-1)) +$$
$$(w - w(n-1))^T \nabla_w^2 J(w(n-1))(w - w(n-1))/2 + \ldots. \qquad (3.3)$$

where

$$\nabla_w^2 J(w(n-1)) = \left(\partial^2 J(w(n-1)) / \partial w_i \partial w_i \right), \; i,j = \overline{1,N} \qquad (3.4)$$

is the matrix of the second order derivatives, also known as the Hessian matrix, or the Hessian. Let us restrict ourselves to a linear approximation of the function $J(w)$: this is possible, if we are not far from the point $w(n\text{-}1)$. In this case the estimate of function $J(w)$ at the point $w(n)$ can be calculated as

$$\tilde{J}(w(n)) = J(w(n\text{-}1)) + (w(n) - w(n\text{-}1))^T \nabla_w J(w(n-1)) \; . \qquad (3.5)$$

The function value at point $w(n)$ will be less than at point $w(n\text{-}1)$ if we choose the new point $w(n)$ from the condition:

$$w(n) - w(n\text{-}1) = -a \; \nabla_w J(w(n-1)) \qquad (3.6)$$

for sufficiently small α. In that case the estimation of the function will be

$$\tilde{J}(w(n)) = J(w(n\text{-}1)) + (w(n) - w(n\text{-}1))^T \nabla_w J(w(n-1)) =$$
$$= J(w(n\text{-}1)) - a \; \nabla_w^T J(w(n-1)) \nabla_w J(w(n-1)) =$$
$$= J(w(n\text{-}1)) - a \left\| \nabla_w J(w(n-1)) \right\|^2 \leq J(w(n\text{-}1)) \; . \qquad (3.7)$$

Inequality (3.7) explains the gradient algorithm that follows from condition (3.6).

3.1.3 Newton's Algorithm

Let us assume that function $J(w)$ is twice differentiable and return again to the Taylor series (3.3). In this case we can restrict ourselves to a quadratic approximation at the point $w(n-1)$:

$$\tilde{J}(w) = J(w(n\text{-}1)) + (w - w(n\text{-}1))^T \nabla_w J(w(n-1)) +$$
$$+ (w - w(n\text{-}1))^T \nabla_w^2 J(w(n-1))(w - w(n\text{-}1))/2 \; .$$

Let us take the minimum point of $\tilde{J}(w)$ as a new approximation of the function $J(w)$. The minimum condition of the function $\tilde{J}(w)$ has the form:

$$\nabla_w \tilde{J}(w) = \nabla_w J(w(n-1)) + \nabla_w^2 J(w(n-1))(w - w(n-1)) = 0. \qquad (3.8)$$

The new approximation follows from (3.8):

$$w(n) = w(n-1) - \left[\nabla_w^2 J(w(n-1))\right]^{-1} \nabla_w J(w(n-1)). \qquad (3.9)$$

Algorithm (3.9) is the Newton's algorithm. In this algorithm the Hessian matrix of the second derivatives is employed instead of the scalar matrix $\gamma(n)I$.

We can also derive this algorithm with a different approach. The minimum point w^* must be a solution of the system of N equations with N variables:

$$\nabla_w J(w) = 0. \qquad (3.10)$$

To solve the system (3.10) let us expand the function $\nabla_w J(w)$ into a Taylor series and restrict ourselves to a linear approximation at point $w(n\text{-}1)$:·

$$\nabla_w \tilde{J}(w) = \nabla_w J(w(n-1)) + \left[\nabla_w^2 J(w(n-1))\right]^{-1}(w - w(n-1)) = 0. \qquad (3.11)$$

Newton's algorithm follows directly from Eq. (3.11).

The rate of convergence of the Newton's algorithm is higher than the rate of convergence of the gradient algorithm. For the quadratic function

$$J(w) = w^T A w / 2 - b^T w \qquad (3.12)$$

with a positive definite symmetric matrix A the Newton's algorithm converges in one step, i.e., $w(1) = w^*$ for any $w(0)$, where $w^* = A^{-1}b$.

This is obvious, since the approximating function $\tilde{J}(w)$ coincides with $J(w)$, the gradient of function $J(w)$ (3.12) is $\nabla_w J(w) = Aw - b$ and the Hessian is equal to $\nabla_w^2 J(w) = A$.

3.1.4 Example Application

There follows an example of the use of the unconstrained minimisation techniques described above when training a Radial Basis Function (RBF) function approximator.

RBF based networks are gaining popularity in the neural networks field as an alternative to the Multi-Layer Perceptron (MLP). They can perform as well as

MLPs in many applications, but can be trained much faster. The basic form of an RBF is:

$$\hat{y}(x) = \sum_{i=1}^{N} a_i \Phi_i \left(\| x - c_i \| \right)$$

Where $a_1, a_2 \dots a_N$ are scalar weights, x is the vector input and c_i, $i=1 \dots N$ are points in the input space called 'centres'. The function Φ can be one of many non-linear functions e.g. multi-quadric, inverse multi-quadric, thin plate spline. The most popular choice for Φ is the gaussian:

$$\Phi_i(r) = e^{-\left(\frac{r}{\sigma_i} \right)^2}$$

Where σ_i is a width parameter associated with the i-th gaussian. In general the Euclidean distance is used as the norm.

Given a set P of example input/output pairs which are observations of some unknown function $y(x)$, the objective is to choose the weights, centres and widths in order to minimise the error, $\varepsilon(x)$, between $y(x)$ and $\hat{y}(x)$ for all members of P.

Defining $J(w)$ as:

$$J(w) = \frac{1}{2} \sum_{P} \varepsilon(x;w)^2$$

We can apply the gradient methods described in this chapter to minimise $J(w)$ and train the RBF. We first need to calculate the partial derivatives of the performance index with respect to the variable parameters. These derivatives are given below:

$$\nabla_{a_k} J(w) = -\sum_{i=1}^{P} \left[e^{-\left(\frac{\| x_i - c_k \|}{\sigma_k} \right)^2} \left(y(x_i) - \sum_{j=1}^{N} a_j e^{-\left(\frac{\| x_i - c_j \|}{\sigma_j} \right)^2} \right) \right]$$

$$\nabla_{c_{kn}} J(w) = -2\sum_{i=1}^{P} \left[\frac{a_k}{\sigma_k^2}(x_{in} - c_{kn}) e^{-\left(\frac{\| x_i - c_k \|}{\sigma_k} \right)^2} \left(y(x_i) - \sum_{j=1}^{N} a_j e^{-\left(\frac{\| x_i - c_j \|}{\sigma_j} \right)^2} \right) \right]$$

$$\nabla_{\sigma_k} J(w) = -2 \sum_{i=1}^{P} \left[a_k \frac{\|x_i - c_k\|^2}{\sigma_k^2} e^{-\left(\frac{\|x_i-c_k\|}{\sigma_k}\right)^2} \left(y(x_i) - \sum_{j=1}^{N} a_j e^{-\left(\frac{\|x_i-c_j\|}{\sigma_j}\right)^2} \right) \right]$$

The parameters can then be updated at each time step, t, according to Eq. (3.2).

$$a_k(t) = a_k(t-1) - \gamma_a \nabla_{a_k} J(w(t-1))$$

$$c_{kn}(t) = c_{kn}(t-1) - \gamma_c \nabla_{c_{kn}} J(w(t-1))$$

$$\sigma_k(t) = \sigma_k(t-1) - \gamma_\sigma \nabla_{\sigma_k} J(w(t-1))$$

Alternatively, the functions can be differentiated again and the Newton method used instead.

3.2 Stochastic Approximation and Recurrent Estimation

3.2.1 Random Input

If the input signal (or the pattern) $u(n)$ is a random variable with probability density function $p(u)$ then for all possible input signals the closeness between the teacher's instruction and the output of the learning system can be given by the performance index $J(w)$, which is an average of $Q(\varepsilon(n;w))$ over U:

$$J(w) = \int_U Q(y^*, y(w)) p(u) du. \tag{3.13}$$

Here it is supposed that $y^*(n) = y^*(n;u)$ and $y(n) = y(n;u,w)$. If we do not know the probability density function $p(u)$ a priori, then we can use an estimate of the function $J(w)$:

$$\tilde{J}(w) = \frac{1}{n} \sum_{m=1}^{n} Q(y^*(m), y(w;m)). \tag{3.14}$$

The optimality conditions for cases (3.13) and (3.14) have the following form:

$$\nabla_w J(w) = \int_U \nabla_w Q(y^*, y(w)) p(u) du = 0, \tag{3.15}$$

$$\nabla_w \tilde{J}(w) = \frac{1}{n} \sum_{m=1}^{n} \nabla_w Q(y^*(m), y(w;m)) = 0, \tag{3.16}$$

respectively.

The approach for solving problems (3.15) and (3.16) is based on the Robbins and Monro stochastic approximation method [2], see also [3].

This method states that, when $\nabla_w J(w)$ is unknown and only a realisation of the random function $\nabla_w Q(y^*, y)$ is available, the solution of problem (3.15) can be achieved using the following recursive procedure

$$w(n) = w(n-1) - \gamma(n) \nabla_w Q(y^*(n), y(w(n-1)), n = 1, 2, \ldots \tag{3.17}$$

where $\gamma(n)$ is a sequence of non-negative real numbers, which has the following properties:

$$\lim_{n \to \infty} \gamma(n) = 0, \sum_{n=1}^{\infty} \gamma(n) = \infty, \sum_{n=1}^{\infty} \gamma^2(n) < \infty . \tag{3.18}$$

For example, the sequence $\gamma(n) = 1/n$ has these properties. The sequence (3.17) converges to the solution of equation (3.15) in some statistical sense.

Algorithm (3.17) is the stochastic alternate version of the gradient algorithm (3.2).

Using the estimate $\nabla_w \tilde{J}(w)$ (3.16) instead of the unknown true value $\nabla_w J(w)$ (3.15), we can calculate the sequence $\gamma(n)$.

For example, let us discuss the simple task of finding the mean value w of the random function u by means of the observations $u(n)$ of function u. The analogues of Eqs. (3.15) and (3.16) are now, respectively

$$\nabla_w J(w) = \int_U (w - u) p(u) du = 0 \tag{3.19}$$

and

$$\nabla_w \tilde{J}(w) = \frac{1}{n} \sum_{m=1}^{n} (w - u(m)) = 0 . \tag{3.20}$$

Algorithm (3.17) has the form

$$w(n) = w(n-1) - \gamma \ (n)(w(n-1) - u(n)) ,$$

where $\gamma (n)$ is constrained by the conditions (3.18).

Using Eq. (3.20) we can directly define the sequence $\gamma (n)$. For this purpose, let us denote the solution of Eq.(3.20) by $w(n)$. The solution after receiving n measurements is given by

$$w(n) = \frac{1}{n}\sum_{m=1}^{n} u(m) . \tag{3.21}$$

Let us transform equation (3.21):

$$w(n) = \frac{1}{n}\sum_{m=1}^{n} u(m) = \frac{1}{n}\left(u(n) + \sum_{m=1}^{n-1} u(m) \right)$$

$$= \frac{1}{n}((n-1)w(n-1) + u(n)) = w(n-1) + \frac{1}{n}(u(n) - w(n-1)) . \tag{3.22}$$

It follows from Eq.(3.22) that in this case the sequence is $\gamma (n) = 1/n$.

In more complicated cases the sequence $\gamma (n)$ can be calculated using the same approach that was used in the derivation of the gradient and Newton's algorithms. For this purpose, let us suppose that for large enough n we have found the solution $w(n)$ of equation (3.16). This implies that:

$$\sum_{m=1}^{n} \nabla_w Q(y^*(m), y(w(n);m)) \equiv 0 . \tag{3.23}$$

For solution $w(n\text{-}1)$ the identity (3.23) has the form

$$\sum_{m=1}^{n-1} \nabla_w Q(y^*(m), y(w(n-1);m)) \equiv 0 . \tag{3.24}$$

We have supposed that n is large enough and, hence, the new n-th measurement will only slightly modify the solution $w(n\text{-}1)$, i. e $\|w(n) - w(n-1)\|$ is small.

Let us expand the function in the left side of identity (3.23) in a Taylor series in the point $w(n-1)$ and restrict ourselves to a linear approximation at this point. This is possible because $\|w(n) - w(n-1)\|$ is small.

$$\sum_{m=1}^{n} \nabla_w Q(y^*(m), y(w(n); m)) = \sum_{m=1}^{n} \nabla_w Q(y^*(m), y(w(n-1); m)) +$$

$$+ \sum_{m=1}^{n} \nabla_w^2 Q(y^*(m), y(w(n-1); m))(w(n) - w(n-1)). \tag{3.25}$$

But according to identity (3.24) only the n-th member in the sum $\sum_{m=1}^{n} \nabla_w Q(y^*(m), y(w(n-1); m))$ is non-zero and Eq. (3.25) becomes the recurrent algorithm

$$w(n) = w(n-1) -$$

$$- \left(\sum_{m=1}^{n} \nabla_w^2 Q(y^*(m), y(w(n-1); m)) \right)^{-1} \nabla_w Q(y^*(m), y(w(n-1); m)). \tag{3.26}$$

Comparing algorithms (3.17) and (3.26), it follows that $\gamma(n)$ can be chosen as a matrix of the form

$$\Gamma(n) = \left(\sum_{m=1}^{n} \nabla_w^2 Q(y^*(m), y(w(n-1); m)) \right)^{-1}. \tag{3.27}$$

Algorithm (3.26) is correct either for large enough n or when $Q(y^*(m), y(w; m))$ is quadratic. In this case the Hessian is not a function of w, the expansion (3.25) is exact, and it is possible to calculate matrix (3.27) on-line. In the opposite case an on-line realisation of algorithm (3.26) is not possible, and we can use only an approximation of matrix $\Gamma(n)$:

$$\tilde{\Gamma}(n) = \left(\sum_{m=1}^{n} \nabla_w^2 Q(y^*(m), y(w(m); m)) \right)^{-1}.$$

It is interesting to note that the recurrent least squares algorithm can be derived using Eq.(3.26).

3.2.2 Example System

The RBF system described in section 3.1.4 updates its parameters after each presentation of the complete training data set (sometimes called 'training by epoch`). It may not always be possible to calculate $J(w)$ if training is taking place on-line, also if very large data sets are used then stochastic algorithms will be necessary. For the RBF case, the derivatives $\nabla_w Q(\hat{y}, y)$ are given by:

$$\nabla_{a_k} Q(\hat{y}, y) = -e^{-\left(\frac{\|x - c_k\|}{\sigma_k}\right)^2}$$

$$\nabla_{c_{kn}} Q(\hat{y}, y) = -2\frac{a_k}{\sigma_k^2}(x_n - c_{kn})e^{-\left(\frac{\|x - c_k\|}{\sigma_k}\right)^2}$$

$$\nabla_{\sigma_k} Q(\hat{y}, y) = -2a_k \frac{\|x - c_k\|}{\sigma_k^3} e^{-\left(\frac{\|x - c_k\|}{\sigma_k}\right)^2}$$

Using Eq. (3.17), the update equations for the parameters at each time step, t, will be:

$$a_k(t) = a_k(t-1) - \gamma_a(t)\nabla_{a_k} Q(\hat{y}, y)$$

$$c_{kn}(t) = c_{kn}(t-1) - \gamma_c(t)\nabla_{c_{kn}} Q(\hat{y}, y)$$

$$\sigma_k(t) = \sigma_k(t-1) - \gamma_\sigma(t)\nabla_{\sigma_k} Q(\hat{y}, y)$$

A suitable $\gamma(t)$ is $\gamma(t)=1/t$.

References

1. B. T. Polyak, *Introduction to optimisation.* New York: Optimization Software, Inc., Publications Division, 1987.
2. H. Robbins, S. Monro "A stochastic approximation method", Annals Math. Stat., 22, 1951, pp. 400-407.
3. M. T. Wasan, *Stochastic approximation.* Cambridge University Press, 1969.

4 Stochastic Algorithms: The Least Squares Method in the Non-Recurrent and Recurrent Forms and the Gauss-Markov Theorem

4.1 The Least Squares Method in Recursive and Non-Recursive Forms (The White Noise Case)

4.1.1 The Least Squares Method in Non-Recursive Form

Assume that the teacher's behaviour is described by the following equation

$$y^*(n) = \sum_{i=1}^{N} w_i^* u_i(n) + \xi(n) = w^{*T} u(n) + \xi(n), \quad n = 1, 2, \ldots \tag{4.1}$$

where vector w^* is unknown to the learning system, whereas $u(n)$ and $y^*(n)$ are available signals to the learning system and $\xi(n)$ is noise. We shall not discuss here the statistical properties of $\xi(n)$ and carry over this problem to the following part of the chapter.

The model for the teacher's instruction in this case is a linear combiner, described by the equation

$$y(n) = \sum_{i=1}^{N} w_i u_i(n) = w^T u(n). \tag{4.2}$$

Let us choose the performance index of the form

$$J(w) = \sum_{m=1}^{n} Q(y^*(m), y(w;m)) ,$$
(4.3)

where

$$Q(y^*(m), y(w;m)) = (y^*(m) - w^T u(m))^2 .$$
(4.4)

The method that provides the minimum for function (4.3)-(4.4) is called the least squares method. The minimum condition for this case is

$$\nabla_w J(w) = \sum_{m=1}^{n} \nabla_w Q(y^*(m), y(w;m)) = -2 \sum_{m=1}^{n} (y^*(m) - w^T u(m)) u(m) = 0$$

or

$$\sum_{m=1}^{n} y^*(m) u(m) = \left(\sum_{m=1}^{n} u(m) u^T(m) \right) w ,$$
(4.5)

from which it follows that the least squares estimate that provides the minimum for function (4.3)-(4.4) is

$$w(n) = \left(\sum_{m=1}^{n} u(m) u^T(m) \right)^{-1} \sum_{m=1}^{n} u(m) y^*(m).$$
(4.6)

The estimate $w(n)$ can be calculated if the matrix $\sum_{m=1}^{n} u(m) u^T(m)$ is invertible, $n \geq N$.

4.1.2 A Priori Information

If we have some *a priori* information about the solution w^* then we can use this information at the first step. For this purpose let us change the function (4.3) to the following form:

$$J^\circ(w) = \alpha \ (w^\circ - w)^T (w^\circ - w) + \sum_{m=1}^{n} (y^*(m) - w^T u(m))^2 \ , \tag{4.7}$$

where $\alpha > 0$ is a suitable constant. The first term in function (4.7) corresponds to a single measurement of w^* with some error v: $w^\circ = w^* + v$.

The minimum condition of the function (4.7) has the form

$$\alpha \ (w^\circ - w) + \sum_{m=1}^{n} (y^*(m) - w^T u(m)) u(m) = 0$$

or

$$\alpha \ w^\circ + \sum_{m=1}^{n} y^*(m) u(m) = \left(\alpha I + \sum_{m=1}^{n} u(m) u^T(m) \right) w \ , \tag{4.8}$$

from which it follows that the least squares estimate with regard to the *a priori* information is:

$$w(n) = \left(\alpha I + \sum_{m=1}^{n} u(m) u^T(m) \right)^{-1} (\alpha \ w^\circ + \sum_{m=1}^{n} y^*(m) u(m)) \ . \tag{4.9}$$

Thus, in this case, the estimate $w(n)$ can be calculated for $n= 1, 2, ...$

Let us find a recursive form for the least squares estimate. For this purpose, let us write the equations for the estimates $w(n)$ and $w(n-1)$ in the case (4.7). The estimate $w(n)$ satisfies the equation

$$\alpha \ w^\circ + \sum_{m=1}^{n} y^*(m) u(m) = \left(\alpha I + \sum_{m=1}^{n} u(m) u^T(m) \right) w(n) \ , \tag{4.10}$$

and the estimate $w(n-1)$ obviously, satisfies the equation

$$\alpha \ w^\circ + \sum_{m=1}^{n-1} y^*(m) u(m) = \left(\alpha I + \sum_{m=1}^{n-1} u(m) u^T(m) \right) w(n-1) \ . \tag{4.11}$$

Now we shall find the relationship between estimates $w(n-1)$ and $w(n)$.
To this end, let us rewrite Eq.(4.11) as follows:

$$\alpha\, w^\circ + \sum_{m=1}^{n-1} y^*(m)u(m)$$

$$= \left(\alpha I + \sum_{m=1}^{n} u(m)u^T(m) \right) w(n-1) - (w(n-1)u^T(n))u(n)$$

(4.12)

Subtracting equation (4.10) from equation (4.12) we obtain

$$\left(\alpha I + \sum_{m=1}^{n} u(m)u^T(m) \right)(w(n) - w(n-1))$$

$$= (y^*(n) - w^T(n-1)u(n))u(n)$$

(4.13)

Denoting the matrix $\left(\alpha I + \sum_{m=1}^{n} u(m)u^T(m) \right)^{-1}$ by $K(n)$, we can rewrite Eq.(4.13) in the recursive form:

$$w(n) = w(n-1) + K(n)(y^*(n) - w^{T}(n-1)u(n))u(n).$$

(4.14)

The Eq.(4.14) relates the estimate $w(n-1)$ obtained at step $(n-1)$ with the estimate $w(n)$. After receiving new measurements $y^*(n), u(n)$ at step n, the estimate $w(n-1)$ is updated on-line. It seems that to update the old estimate $w(n-1)$ using Eq.(4.14), it is necessary to invert matrix $K^{-1}(n)$.

This direct matrix inversion can be overcome using the following relation that connects two invertible $N \times N$-matrixes A and B and two N-dimensional vectors u and v, known as the matrix inversion lemma. If

$$A = B + u v^T ,$$

(4.15)

then

$$A^{-1} = B^{-1} - \frac{B^{-1}u\, v^T B^{-1}}{1 + v^T B^{-1} u}.$$

(4.16)

This can be easy verified by multiplying the right sides of Eqs. (4.15) and (4.16). If we use relation (4.16) for matrix

$$K^{-1}(n) = \alpha\, I + \sum_{m=1}^{n} u(m)u^{T}(m) = K^{-1}(n-1) + u(n)u^{T}(n) \qquad (4.17)$$

then we get

$$K(n) = K(n-1) - \frac{K(n-1)u(n)u^{T}(n)K(n-1)}{1 + u^{T}(n)K(n-1)u(n)}. \qquad (4.18)$$

Eq.(4.14) together with Eq.(4.18) form the recursive least squares algorithm.

It should be pointed out that the structure of the recursive algorithm is not dependent on the *a priori* information. But the presence of the *a priori* information gives us the possibility to begin the recursive process from the first point. In this case $w(0) = w^{\circ}$ and $K(0) = \alpha^{-1} I$.

4.2 The Gauss-Markov Theorem

4.2.1 Optimal Estimates

Teacher's behaviour: In this part which is based on [1], [3] it is supposed that the teacher's behaviour is described by the following equation

$$y^{*}(m) = \sum_{i=1}^{N} w_i^{*} u_i(m) + \xi(m) = w^{*T} u(m) + \xi(m), \quad m = 1, 2, \ldots, n \quad (4.19)$$

where vector w^{*} is unknown to the learning system, input vector $u(m)$ and teacher's instruction $y^{*}(m)$ are available signals for the learning system, $u(m), w^{*} \in R^{N}$. The correlation between $u(i)$ and $\xi(j)\ \forall\ i, j$ is zero ($E\{u(i)\xi(j)\} = 0\ \forall\ i, j$). The statistical properties of the noise $\xi(m)$ with zero mean value ($E\{\xi(m)\} = 0$) are described by means of the $n \cdot n$ covariance matrix

$$D_\xi(n) = \begin{bmatrix} d_{11} & d_{12} & \cdot & d_{1n} \\ d_{21} & d_{22} & \cdot & d_{21} \\ \cdot & \cdot & \cdot & \cdot \\ d_{n1} & d_{n2} & \cdot & d_{nn} \end{bmatrix} = E\left\{\Xi(n)\Xi^T(n)\right\}, \tag{4.20}$$

where E is the symbol representing statistical expectation,

$$\Xi^T(n) = (\xi(1), \xi(2), ..., \xi(n)) \tag{4.21}$$

is a vector of the noise components and $d_{ij} = E\left\{\xi(i)\xi(j)\right\}$.

Let

$$Y^{*T}(n) = (y^*(1), y^*(2), ..., y^*(n)) \tag{4.22}$$

be a vector of the teacher's instruction and

$$U^T(n) = (u(1), u(2), ..., u(n)) \tag{4.23}$$

be a matrix of the input signals $u(m)$.

Using notation (4.21)-(4.23) we can rewrite equations (4.19) in the form

$$Y^*(n) = U(n)w^* + \Xi(n). \tag{4.24}$$

The Search For An Estimator: Let us construct the estimate $w(n)$ of the unknown vector w^* satisfying the following conditions:

1. The estimate $w(n)$ is linear combination of the teacher's instructions:
$$w(n) = X Y^*(n), \tag{4.25}$$
where X is unknown $N \cdot n$ - matrix,
2. The estimate $w(n)$ is unbiased:
$$E\left\{w(n)\right\} = w^*, \tag{4.26}$$
3. The variance of the estimate $w(n)$ is minimal:
$$\min_X E\left\{(w(n) - w^*)^T (w(n) - w^*)\right\}. \tag{4.27}$$

The variance of the estimate $w(n)$ is a sum of the variances of the each components of the vector $w(n)$. If an estimate satisfying Eqs. (4.1)-(4.4) exists, it is called the Gauss-Markov estimate.

Determination of The Gauss-Markov Estimate: Using Eq.(4.24) let us rewrite the conditions (4.25) and (4.26):

$$w(n) = X Y^*(n) = X(U(n)w^* + \Xi(n)) = X U(n)w^* + X \Xi (n),\qquad(4.28)$$

$$E\{w(n)\} = E\{X U(n)w^* + X \Xi(n)\} = X U(n)w^* = w^*.\qquad(4.29)$$

The expression (4.29) follows from condition (4.26). From Eqs.(4.28) and (4.29) we have

$$w(n) - w^* = X \Xi (n).\qquad(4.30)$$

Thus, the problem of finding the linear unbiased estimate $w(n)$ with minimum variance, Eqs.(4.25)-(4.27), is reduced to the task of the constrained minimisation:

$$\min_{X} E\{\Xi(n)^T X^T X \Xi(n)\},\qquad(4.31)$$

$$X U(n)w^* = w^*.\qquad(4.32)$$

Let us consider the condition (4.31). It is evident that,

$$Tr\, A = Tr\, A^T, \quad Tr(AB) = Tr(BA), \quad a^T a = Tr(aa^T),$$

where $Tr\, A = \displaystyle\sum_{i=1}^{N} a_{ii}$.

Hence

$$E\{\Xi^T (n)X^T X \Xi(n)\} = Tr\, E\{X \Xi(n)\Xi(n)^T X^T\} = Tr(X D_\xi(n) X^T)$$

and the condition (4.31) takes the form

$$\min_{X} Tr(X D_\xi(n) X^T).\qquad(4.33)$$

The Eq. (4.32) must be fulfilled for any vector w^* and hence,

$$X U(n) = I.\qquad(4.34)$$

To solve the constrained minimum task (4.33)-(4.34) let us introduce the Lagrange function

$$L(X) = Tr(XD_\xi(n)X^T) + Tr(\Lambda(XU(n) - I)),$$ (4.35)

which consists of the objective function $Tr(XD_\xi(n)X^T)$ and constraints multiplied by the $N \cdot N$ Lagrange matrix. Now we have the unconstrained minimum problem:

$$\min_X L(X) = \min_X (Tr(XD_\xi(n)X^T) + Tr(\Lambda(XU(n) - I))).$$ (4.36).

It is possible to show that for any square matrix G there are the following rules of differentiation of the trace forms:

$$\frac{\partial\, Tr(XGX^T)}{\partial X} = XG + XG^T,$$ (4.37)

$$\frac{\partial\, Tr(XG)}{\partial X} = G^T,$$ (4.38)

from which follows the minimum condition of the Lagrange function (4.35):

$$\nabla_X L(X) = 2X\, D_\xi(n) + \Lambda^T U^T(n) = 0,$$ (4.39)

$$\nabla_\Lambda L(X) = XU(n) - I = 0.$$ (4.40)

In order to solve the system of linear equations (4.39), (4.40), let us find the matrix X from Eq.(4.39) as a function of the matrix Λ :

$$X = -\frac{1}{2}\Lambda^T U^T(n)D_\xi^{-1}(n).$$ (4.41)

Substituting (4.41) into Eq.(4.40) we can find the matrix Λ^T :

$$\Lambda^T = -2(U^T(n)D_\xi^{-1}(n)U(n))^{-1}.$$ (4.42)

Now substituting Λ^T (4.42) into Eq. (4.41) we find the solution X :

$$X(n) = (U^T(n)D_\xi^{-1}(n)U(n))^{-1}U^T(n)D_\xi^{-1}(n). \tag{4.43}$$

Substituting the solution (4.43) into Eq.(4.25) we can find the *optimal Gauss-Markov estimate w(n) of the vector* w^* :

$$w(n) = (U^T(n)D_\xi^{-1}(n)U(n))^{-1}U^T(n)D_\xi^{-1}(n)Y^*(n). \tag{4.44}$$

as an linear unbiased estimate with the minimum variance. It is easy to show using properties (4.37) and (4.38) that the estimate (4.44) minimises the function

$$J^\circ(w) = (Y^*(n) - U(n)w)^T D_\xi^{-1}(n)(Y^*(n) - U(n)w). \tag{4.45}$$

It means that the estimate (4.44) is the solution of the equation $\nabla_w J^\circ(w) = 0$.

From Eqs. (4.30) and (4.41) it follows, that the covariance matrix of the estimate $w(n)$ is equal to

$$P(n) = E\left\{(w(n) - w^*)(w(n) - w^*)^T\right\} = (U^T(n)D_\xi^{-1}(n)U(n))^{-1}.$$

The function $J^\circ(w)$ is the generalised case of the usual least squares function when the noise $\xi(m)$ is white with the covariance matrix $D_\xi(n) = \sigma_\xi^2 I(n)$, where σ_ξ^2 is the variance of the noise and $I(n)$ is $n \cdot n$ unit matrix. In this case the function (4.45) takes the form

$$J^\circ(w) = \sigma_\xi^{-2}(Y^*(n) - U(n)w)^T(Y^*(n) - U(n)w)$$

$$= \sigma_\xi^{-2}\sum_{m=1}^{n}(y^*(m) - \sum_{i=1}^{N} w_i\ u_i(m))^2$$

and coincides with the usual least squares function. Thus, the Gauss-Markov estimates are the generalised least squares estimates and the latter are optimal only if the noise $\xi(n)$ is white. *The Gauss-Markov theorem asserts that estimate (4.44) has the smallest variance among all linear unbiased estimates independently of the noise distribution law.*

By the deriving the Gauss-Markov estimates it was implicitly supposed that the matrixes $D_\xi(n)$ and $U^T(n)D_\xi^{-1}(n)U(n)$ are invertible. If it is not the case than we must use another approach, for example the pseudo-inverse method. We will not discuss this situation here.

4.2.2 Connection With the Maximum Likelihood Estimates [2]

Let us suppose that the teacher's instructions are described by the same equations, (4.19) and (4.20), and that the noise has a Gaussian distribution law. In this case the *a posteriori* density function as it follows from the probability theory, has the form

$$p_{a\ post}(w) = const.\exp(-(Y^*(n) - U(n)w)^T D_\xi^{-1}(n)(Y^*(n) - U(n)w)).$$

The performance index in the maximum likelihood method is constructed by taking the logarithm of the a posteriori probability density function and, hence, in our situation when the noise has the Gaussian distribution law, the performance index $J(w) = -\ln p_{a\ post}(w)$ and hence, coincides with the function $J^\circ(w)$ (4.27). It follows that the Gauss-Markov estimate coincides with the maximum likelihood estimate in the case when the noise has a Gaussian distribution

Let us suppose, that memory w is random and we know that the *a priori* density function $p_{a\ pri}(w)$ is Gaussian with the mean value w° and the covariance matrix D_w. We can write the *a posteriori* probability density function now in the form:

$$const.\exp(-(w^\circ - w)^T D_w^{-1}(w^\circ - w)$$
$$-(Y^*(n) - U(n)w)^T D_\xi^{-1}(n)(Y^*(n) - U(n)w))$$

and hence the performance index in this case is equal to

$$J(w) = (w^\circ - w)^T D_w^{-1}(w^\circ - w) + (Y^*(n) - U(n)w)^T D_\xi^{-1}(n)(Y^*(n) - U(n)w).$$

We can get the maximum likelihood estimate from the condition

$$w^\circ \ \nabla_w J(w) = -2D_w^{-1}(w^\circ - w) - 2U^T(n)D_\xi^{-1}(n)(Y^*(n) - U(n)w) = 0,$$

where the *a priori* information on the w° and D_w is taken into account in the form:

$$w(n) = (D_w^{-1} + U^T(n)D_\xi^{-1}(n)U(n))^{-1}(D_w^{-1}w^\circ + U^T(n)D_\xi^{-1}(n)Y^*(n)).$$

Usually, when n tends to infinity the influence of w° and D_w on the solution tends to zero.

4.3 Example System

In the previous chapter, Radial Basis Functions (RBFs) were introduced as an example application for unconstrained minimisation techniques. In practice, however, it is more common to fix the RBF centres and widths using heuristic methods (e.g. k-means clustering for centre selection). In this case, training the RBF becomes a linear problem. Hence RBFs can be used in applications (e.g. adaptive control) where the training time of a multi-layer perceptron using the back-propagation algorithm would be prohibitively large.

Recalling that the form of an RBF is given by:

$$\hat{y}(x) = \sum_{i=1}^{N} a_i \Phi_i \left(\| x - c_i \| \right)$$

and we have P input/output examples. Then if the centres and widths are fixed, we need to solve for a (the vector of weights $a_1 \ldots a_N$) the following over-determined set of linear simultaneous equations in the least squares sense:

$$Y = \Theta a$$

where Y is a vector of the P desired outputs and Θ has the form:

$$\Theta = \begin{bmatrix} \Phi(\|x_1 - c_1\|) & \Phi(\|x_1 - c_2\|) & \cdots & \Phi(\|x_1 - c_N\|) \\ \Phi(\|x_2 - c_1\|) & \Phi(\|x_2 - c_2\|) & \cdots & \Phi(\|x_2 - c_N\|) \\ \vdots & \vdots & \vdots & \vdots \\ \Phi(\|x_P - c_1\|) & \Phi(\|x_P - c_2\|) & \cdots & \Phi(\|x_P - c_N\|) \end{bmatrix}$$

This equation can be solved in a number of different ways. Using Eq. (4.6), the least squares estimate is given by:

$$a = \Theta^{-1} Y$$

Alternatively, the recursive form of the least squares estimate can be used:

$$a(n) = a(n-1) + K(n)\left(y(n) - a^T(n-1)u(n)\right)u(n)$$

$$K(n) = K(n-1) - \frac{K(n-1)u(n)u^T(n)K(n-1)}{1 + u^T(n)K(n-1)u(n)}$$

where $u(n)$ is the n-th row of Θ.

References

1. Aved'yan E. D., *Recurrent methods of signal processing*. Moscow; IPK MRP, 1986 (in Russian).
2. Hamilton W., *Statistics in physical science*. New York, 1964.
3. Brammer K., Siffling G., *Kalman-Bucy-Filter*. München, Wien: Oldenburg Verlag, 1980.

5 Stochastic Algorithms

5.1 Algorithms With Forgetting Factor

For non-stationary white noise $\xi(n)$ the covariance matrix D_ξ is diagonal with elements $d_{m,m} = \sigma_\xi^2(m)$. The minimising function (see Eq. (4.45), Lecture 4) takes the form

$$J^\circ(w) = \sum_{m=1}^{n} (y^*(m) - w^T u(m))^2 \sigma_\xi^{-2}(m) \tag{5.1}$$

where $\sigma_\xi^{-2}(m)$ plays the role of weights in the sum (5.1). The estimate $w(n)$ that provides the minimum for function (5.1) satisfies the equation

$$\sum_{m=1}^{n} (y^*(m) - w^T(n)u(m))u(m)\sigma_\xi^{-2}(m) = 0 \tag{5.2}$$

and the estimate $w(n-1)$, obviously, satisfies the equation

$$\sum_{m=1}^{n-1} (y^*(m) - w^T(n-1)u(m))u(m)\sigma_\xi^{-2}(m) = 0. \tag{5.3}$$

Now we shall find the connection between estimates $w(n-1)$ and $w(n)$. To this end let us rewrite the Eq. (5.3) in the form:

$$\sum_{m=1}^{n} (y^*(m) - w^T(n-1)u(m))u(m)\sigma_\xi^{-2}(m) - (y^*(n)$$

$$- w^T(n-1)u(n))u(n)\sigma_\xi^{-2}(n) = 0$$

(5.4)

Multiplying Eqs. (5.3) and (5.4) by $\sigma_\xi^2(n)$ and subtracting obtained equations we have

$$\sigma_\xi^2(n)\sum_{m=1}^{n} \sigma_\xi^{-2}(m)u(m)u^T(m)(w(n) - w(n-1))$$

$$- (y^*(n) - w^T(n-1)u(n))u(n) = 0$$

or

$$w(n) - w(n-1) =$$

$$\left(\sigma_\xi^2(n)\sum_{m=1}^{n} \sigma_\xi^{-2}(m)u(m)u^T(m)\right)^{-1} (y^*(n) - w^T(n-1)u(n))u(n)$$

(5.5)

Let us denote the matrix $\left(\sigma_\xi^2(n)\sum_{m=1}^{n} \sigma_\xi^{-2}(m)u(m)u^T(m)\right)^{-1}$ by $K(n)$, then we can rewrite the Eq. (5.5) in the recurrent form:

$$w(n) = w(n-1) + K(n)(y^*(n) - w^T(n-1)u(n))u(n).$$

(5.6)

Equation (5.6) connects the estimate $w(n-1)$ obtained at step $(n-1)$ with the estimate $w(n)$ and allows the possibility of on-line calculation: i.e. recalculating the estimate $w(n-1)$ after receiving, at step n, the new measurements $y^*(n), u(n)$. To recalculate the old estimate $w(n-1)$ using Eq.(5.6) it appears necessary to invert matrix $K^{-1}(n)$.

Direct matrix inversion can be bypassed using the relation that connect two invertible $N \cdot N$ matrixes A and B and N dimensional vectors u and v (we have used this relation in the previous lecture). If

$$A = B + u v^T,$$

(5.7)

then

$$A^{-1} = B^{-1} - \frac{B^{-1}u\,v^T B^{-1}}{1+uv^T B^{-1}}. \tag{5.8}$$

This can be easily verified by multiplying the right sides of Eqs. (5.7) and (5.8). This is known as the matrix inversion lemma. If we use the relation (5.8) for matrix

$$
\begin{aligned}
K^{-1}(n) &= \sigma_\xi^2(n) \sum_{m=1}^{n} \sigma_\xi^{-2}(m)u(m)u^T(m) \\
&= \sigma_\xi^2(n) \sum_{m=1}^{n-1} \sigma_\xi^{-2}(m)u(m)u^T(m) + u(n)u^T(n) \\
&= \frac{\sigma_\xi^2(n)}{\sigma_\xi^2(n-1)} \left(\sigma_\xi^2(n-1) \sum_{m=1}^{n} \sigma_\xi^{-2}(m)u(m)u^T(m) \right) + u(n)u^T(n) \\
&= \frac{\sigma_\xi^2(n)}{\sigma_\xi^2(n-1)} K^{-1}(n-1) + u(n)u^T(n)
\end{aligned}
\tag{5.9}
$$

then we obtain

$$K(n) = \mu^{-1}(n)\left(K(n-1) - \frac{K(n-1)u(n)u^T(n)K(n-1)}{\mu(n)+u^T(n)K(n-1)u(n)} \right), \tag{5.10}$$

where $\mu(n) = \sigma_\xi^2(n)/\sigma_\xi^2(n-1)$.

If $\mu(n) < 1$, then the algorithm (5.6), (5.10) can be called the recurrent algorithm with forgetting factor $\mu(n)$ because old information in function (5.1) has less significance than new information when calculating the estimates of the system parameters. It gives the algorithm a tracking capability. In the particular case when

$$\sigma_\xi^2(n) = \sigma_\xi^2 \mu^n, \ 0 < \mu < 1, \tag{5.11}$$

the forgetting factor $\mu(n)$ in the algorithm (5.10) is constant and equal to μ.

5.2 The Least Squares Method by Correlated Noise in the Non-Recursive and Recursive Forms: Connection With Decorrelation Procedures

5.2.1 The Arbitrary Case

In the previous lecture we have derived the Gauss-Markov estimate which is optimal in the presence of correlated noise (in this case the matrix D is not diagonal). Let us find now the corresponding recursive form of this estimate.

Using the notation from the Lecture 4, the estimates $w(n)$ and $w(n-1)$ are determined from the equation

$$U^T(n)D_\xi^{-1}(n)U(n)\,w(n) = U^T(n)D_\xi^{-1}(n)Y^*(n), \tag{5.12}$$

$$
\begin{aligned}
U^T(n-1)D_\xi^{-1}&(n-1)U(n-1)\,w(n-1) \\
&= U^T(n-1)D_\xi^{-1}(n-1)Y^*(n-1)
\end{aligned} \tag{5.13}
$$

The matrix $D_\xi(n)$ is formed by bordering the matrix $D_\xi(n-1)$:

$$D_\xi(n) = \begin{bmatrix} D_\xi(n-1) & d(n-1) \\ d^T(n-1) & d_{nn} \end{bmatrix}, \tag{5.14}$$

where

$$d^T(n-1) = (d_{n,1},\ldots,d_{n,n-1}) = E\left\{\xi(n)\Xi^T(n-1)\right\}. \tag{5.15}$$

It is possible to verify by direct multiplication of the block matrices $D_\xi(n)$ and $D_\xi^{-1}(n)$, that the matrix $D_\xi^{-1}(n)$ has the following representation:

$$D_\xi^{-1}(n) = \begin{bmatrix} B_\xi(n-1) & \dfrac{-D_\xi^{-1}(n-1)d(n-1)}{\alpha(n)} \\ \dfrac{-d^T(n-1)D_\xi^{-1}(n-1)}{\alpha(n)} & \alpha^{-1}(n) \end{bmatrix}, \tag{5.16}$$

where matrix $B_\xi(n-1)$ and parameter $\alpha(n)$ have the form:

$$B_\xi(n-1) = D_\xi^{-1}(n-1) + \frac{D_\xi^{-1}(n-1)d(n-1)d^T(n-1)D_\xi^{-1}(n-1)}{\alpha(n)}, \qquad (5.17)$$

$$\alpha(n) = d_{nn} - d^T(n-1)D_\xi^{-1}(n-1)d(n-1). \qquad (5.18)$$

Taking into account the fact that the matrix $U^T(n) = (U^T(n-1), u(n))$ and the matrices. $D_\xi^{-1}(n)$ and $D_\xi^{-1}(n-1)$ are related by Eq. (5.16), it may be shown that

$$K^{-1}(n) = K^{-1}(n-1) + u_{**}(n)u_{**}^T(n), \qquad (5.19)$$

$$K^{-1}(n) = U^T(n)D_\xi^{-1}(n)U(n), \qquad (5.20)$$

where

$$u_{**}(n) = u_*(n)/\alpha^{1/2}(n), \qquad (5.21)$$

$$u_*(n) = u(n) - U^T(n-1)D_\xi^{-1}(n-1)d(n-1). \qquad (5.22)$$

Using analogous transformations, it may be shown that

$$U^T(n)D_\xi^{-1}(n)Y^*(n) = U^T(n-1)D_\xi^{-1}(n-1)Y^*(n-1) + u_{**}(n)y_{**}^*(n), \quad (5.23)$$

$$y_{**}^*(n) = y_*^*(n)/\alpha^{1/2}(n), \qquad (5.24)$$

$$y_*(n) = y(n) - Y^{*T}(n-1)D_\xi^{-1}(n-1)d(n-1). \qquad (5.25)$$

Let us add $(u_{**}(n)u_{**}^T(n))w(n-1)$ to both sides of Eq. (5.13) and subtract the corresponding parts of the transformed equation (5.13) from the right and left sides of Eq.(5.12). Then with allowance for (5.19) and (5.23) we obtain the relationship

$$K^{-1}(n)(w(n) - w(n-1)) = (y_{**}^*(n) - w^T(n-1)u_{**}(n))u_{**}(n) \qquad (5.26)$$

or

$$w(n) = w(n-1) + K(n)(y_{**}^*(n) - w^{T}(n-1)u_{**}(n))u_{**}(n). \tag{5.27}$$

Taking into account Eq. (5.19) and matrix inversion lemma we also obtain the recurrent relationship for the matrix $K(n)$:

$$K(n) = K(n-1) - \frac{K(n-1)u_{**}(n)u_{**}^{T}(n)K(n-1)}{1 + u_{**}^{T}(n)K(n-1)u_{**}(n)}. \tag{5.28}$$

The derived relations (5.27) and (5.28) form an algorithm for estimating the vector in the presence of correlated noise. The algorithm differs from the recursive algorithm for the white noise case, given in the previous lecture, in that instead of the measurements $y^*(n)$ and vectors $u(n)$, the transformed quantities $y_{**}^*(n)$, $u_{**}(n)$ determined from Eq.(5.21) and (5.24) appear in the derived algorithm. These quantities are linear combinations of the observations $y^*(m)$ and vectors $u(m), m = \overline{1,n}$, respectively. It is possible to show [2], that the random component of the transformed process $y_{**}^*(n)$ is discrete white noise and parameter $\alpha(n)$ is equal to the variance of the process $y_*^*(n)$.

All of this allows the conclusion that the procedure for obtaining $y_{**}^*(n)$ corresponds to the linear transformation L of the random vector

$$Y^*(n) = U(n)w + \Xi(n):$$
$$L(Y^*(n)) = L(U(n))w + L(\Xi(n)) \tag{5.29}$$

which has non-correlated components with unit variance:

$$E\left\{L(\Xi(n))L^{T}(\Xi(n))\right\} = I.$$

The procedure for obtaining $y_{**}^*(n), u_{**}(n)$ given by (5.21), (5.22) and (5.24), (5.25) is valid for an arbitrary covariance matrix $D_\xi(n)$ and therefore is very complicated: it is necessary invert the $n \cdot n$ matrixes with growing n.

5.2.2 Special Case

Let us suppose that the correlated noise $\xi(n)$ in the observations $y^*(n) = w^{*T} u(n) + \xi(n)$ is a stationary random process formed by the passage of discrete white noise $\eta(n)$ through a filter having a known transfer function $W(z) = P(z)/Q(z)$, $P(z)$ and $Q(z)$ are polynomials on z, where z is the shift operator: $z(y(n)) = y(n+1)$.

In paper [2] results are given which state that the decorrelation operator L for obtaining $y_{**}^*(n)$, $u_{**}(n)$ is realised by the passage of the observed processes through a filter with varying parameters. This filter converges to a stable filter with constant parameters. Since the determination of the transfer function of such a filter is accomplished fairly simply, while the calculation of the processes $y_{**}^*(n)$, $u_{**}(n)$ is fairly cumbersome, one may reject the precise procedure for formulating $y_{**}^*(n)$, $u_{**}(n)$, given in [2] and use the processes at the output of the filter having transfer function

$$W_*(z) = W^{-1}(z) = Q(z)/P(z), \tag{5.30}$$

if the transfer function $W(z)$ is minimum phase and

$$W_*(z) = Q(z)/P^*(z), \tag{5.31}$$

if the transfer function is non-minimum phase. It means that the same absolute values of the zeros z_v of the polynomial $P(z)$ are greater than unity: $|z_v| > 1$. In this case the order of the polynomial $P^*(z)$ is equal to the order of the polynomial $P(z)$. The zeros of $P^*(z)$ are constructed using the following rule: zeros of $P^*(z)$ coincide with zeros z_v of $P(z)$ if $|z_v| < 1$. Alternatively if the absolute value z_v is greater than one, $|z_v| > 1$, they are equal to z_v^{-1}:

$$z_v^* = \begin{cases} z_v, & if\ |z_v| < 1 \\ z_v^{-1}, & if\ |z_v| > 1 \end{cases}. \tag{5.32}$$

It is easy to show that if $z = e^{jw}$ then $|P(z)/P^*(z)| = const.$.

5.3 Introduction to the Kalman filter [1], [3]

Until now we have considered different kinds of algorithms for the situation when the characteristics of the teacher's instructions were not time varying or they were not directly taken into account.

Let us suppose now, that the memory w^* in the teacher's instructions $y^*(n)$ is time varying, we know the law of the variation of the memory in implicit form given by the linear difference equation

$$w^*(n) = A(n)w^*(n-1) + B(n)v(n) + C(n)\mu(n),$$
(5.33)

where known matrices $A(n)$, $B(n)$, $C(n)$, have dimensions $N \cdot N$, $N \cdot M$, and $N \cdot L$, respectively, $v(n)$ is a M-dimensional known vector process, and $\mu(n)$ is L-dimensional white noise vector process with the known covariance matrix $Q(m) = E\{\mu(n)\mu^T(n)\}$. The teacher's K-dimensional vector instructions have the form

$$y^*(n) = U(n)w^*(n) + \xi(n),$$
(5.34)

where $U(n)$ is a known $K \cdot N$ matrix function of the time and $\xi(n)$ is unknown white noise with the covariance matrix $R(n)$. The processes $\mu(n)$ and $\xi(n)$ are uncorrelated. The model of teacher's instructions (5.33), (5.34) generalises the simple static model of teacher's instructions, which we used earlier.

Let us chose a learning model having the same structure as the teacher's instructions:

$$y(n) = U(n)w(n).$$
(5.35)

In this case the problem of adjusting of the model's parameters becomes a problem of optimal estimation of the stochastic process $w^*(n)$.

If we can estimate process $w^*(n)$ with minimum variance then we can use the Gauss-Markov theorem.

Let us suppose that we have find the best estimate of $w(n)$, in the Gauss-Markov sense, using $w(n-1)$ and observations of $U(n)$ and $y^*(n)$. This estimate may be written as

$$w(n-1) = w^*(n-1) + \Delta\, w(n-1),$$
(5.36)

where $\Delta w(n-1)$ is a random vector-error with the unknown covariance matrix

$$P(n-1) = E\left\{\Delta(n-1)\Delta^{T}(n-1)\right\}.$$
(5.37)

Multiply both sides of Eq. (5.36) by the matrix $A(n)$ and using Eq.(5.33) we can write

$$A(n)w(n-1) + B(n)v(n) = w^*(n) - C(n)\mu(n) + A(n)\Delta w(n-1),$$
(5.38)

where on the left side is the known vector $A(n)w(n-1) + B(n)v(n)$, and on the right side is the unknown vector $w^*(n)$ and random unknown vector

$$\varsigma(n) = -C(n)\mu(n) + A(n)\Delta w(n-1).$$
(5.39)

We can combine the equation (5.34) with (5.39)

$$\begin{pmatrix} A(n)w(n-1) + B(n)v(n) \\ y^*(n) \end{pmatrix} = \begin{pmatrix} I \\ U(n) \end{pmatrix} w^*(n) + \begin{pmatrix} \varsigma(n) \\ \xi(n) \end{pmatrix}.$$
(5.40)

If we introduce the notation

$$\left. \begin{aligned} y_{o}(n) &= \begin{pmatrix} A(n)w(n-1) + B(n)v(n) \\ y^*(n) \end{pmatrix} \\ U_{o}(n) &= \begin{pmatrix} I \\ U(n) \end{pmatrix} \\ \xi_{o}(n) &= \begin{pmatrix} \varsigma(n) \\ \xi(n) \end{pmatrix} \end{aligned} \right\},$$
(5.41)

then we can rewrite (5.40) in the usual form

$$y_{o}(n) = U_{o}(n)w^*(n) + \xi_{o}(n),$$
(5.42)

where the noise $\xi_{o}(n)$ has the covariance matrix

$$D_{\xi_o}(n) = E\left\{\xi_o(n)\xi_o{}^T(n)\right\} = E\left\{\begin{pmatrix}\varsigma(n)\\\xi(n)\end{pmatrix}\left(\varsigma^T(n),\xi^T(n)\right)\right\}.\qquad(5.43)$$

Using the expression for the optimal Gauss-Markov estimate with respect to the system (5.42)

$$w(n) = \left(U_o^T(n)D_{\xi_o}^{-1}(n)U_o(n)\right)^{-1}U_o^T(n)D_{\xi_o}^{-1}(n)y_o(n),$$

expressions for the $U_o(n), y_o(n)$, Eqs.(5.41) and the expression for the matrix $D_{\xi_o}(n)$, Eq.(5.43) we can derive the recursive algorithm for the Kalman filter, which makes possible the calculation of the new estimate from the old $w(n-1)$.

References

1. Avedyan E. D., *Recurrent methods of signal processing*. Moscow; IPK MRP, 1986 (in Russian).
2 Avedyan E. D. ,"Recurrent method of least squares in the presence of correlated interferences", Automation and remote control, No. 5, pp. 760-768, 1975.
3. Brammer K., Siffling G., *Kalman-Bucy-Filter*. München, Wien: Oldenburg Verlag, 1980.

6 Multilayer Neural Networks

6.1 Multilayer Neural Network as a Non-Linear Transformer. The Kolmogorov and Cybenko Theorems.

6.1.1 Introduction

Multilayer neural networks are typical learning systems. Here we can find all the necessary components of a learning system: a performance index, a memory, and learning algorithms. Being designed according to the principles of their biological analogues, multilayer neural networks (MNN) are able to solve a wide range of problems in pattern recognition [1], identification [2], control of complex dynamical non-linear systems [3], [4], robot control [5], etc.

The main analogy between a MNN and the biological nervous system is that each consists of a large number of simple parallel processing elements, these elements form a network, and this network is able to learn and solve complicated problems.

A strong impulse for much of the subsequent research in the field of MNNs was given in 1986 when the "Parallel Distributed Processing" (PDP) group published a series of results and algorithms concerning multilayer neural networks. In one of these publications [6] the back-propagation algorithm was implemented by Rumelhart, Hinton and Williams as a method for calculating the gradient of the function which is the performance index for this system. The back-propagation algorithm was first derived by Werbos in 1974 but, unfortunately, Werbos's work remained unknown until 1986.

The learning process of multilayer neural networks is carried out by comparison of the network's outputs. (i.e., the outputs of the neurons of the last layer) with the corresponding teacher's commands. Information about the ideal outputs of the neurons of the hidden layers is absent. Despite this, the knowledge of the structure of an MNN allows us to calculate the correcting signals not only for the neurons of the last layer but for the other layers too.

6.1.2 MNN Architecture

The basic processing element in an multilayer neural network is a non-linear combiner (introduced in the first chapter) called a neuron by analogy with neurobiology. Such a network (Fig. 6.1) is characterised by the following: the network consists of M layers, each μ-th layer is formed by N_μ neurons, there are no connections between neurons of the same layer. The layers between the input vector and the output layer are named hidden layers. Notice that the first layer is also a hidden layer if the number of layers is greater than one. The neurons are fully connected between layers, i.e. outputs of neurons in the μ -th layer are fed to the input of every neuron in the following (μ +1)-th layer. The external input signal is fed to the inputs of the neurons of the first layer, and the outputs of the neurons in the last (M-th) layer form the output vector of the network.

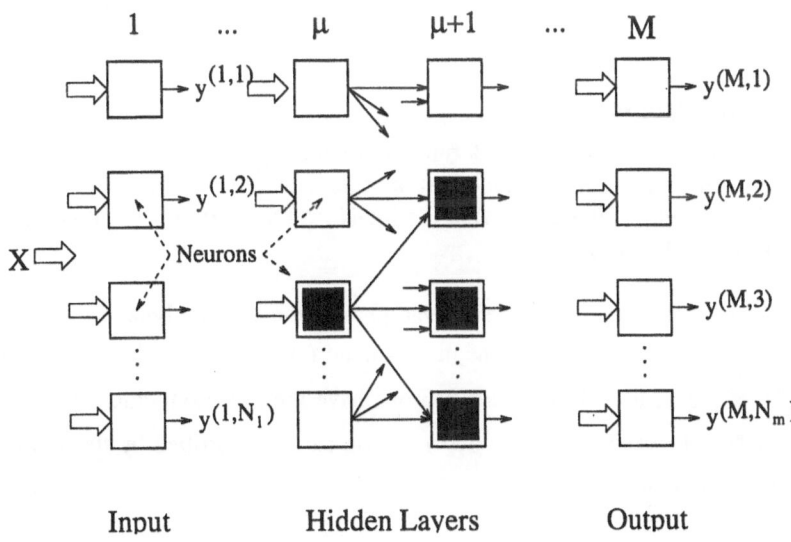

Fig 6.1 Topology of a Neural Network

The ν-th neuron of the μ-th layer ((μ,ν)-neuron) may be considered to have two components: the combiner and the activation function.

The Combiner: The combiner, or discriminant function of *order one:*

$$net_1^{(\mu,\nu)} = w_0^{(\mu,\nu)} + \sum_{i=1}^{N} w_i^{(\mu,\nu)} x_i^{(\mu,\nu)} = w^{T(\mu,\nu)} u^{(\mu,\nu)} \tag{6.1}$$

where $w^{(\mu,\nu)} = (w_0^{(\mu,\nu)}, w_1^{(\mu,\nu)}, ..., w_N^{(\mu,\nu)})^T$ is the weight vector, $u^{(\mu,\nu)} = (1, u_1^{(\mu,\nu)}, ..., u_N^{(\mu,\nu)})^T$ is the extended input vector of the neuron, and $x_i^{(\mu,\nu)}$ is the i-th component of the N-dimensional input vector $x^{(\mu,\nu)}$.

The combiner, or discriminant function of *order two*:

$$net_2^{(\mu,\nu)} = w_0^{(\mu,\nu)} + \sum_{i=1}^{N} w_i^{(\mu,\nu)} x_i^{(\mu,\nu)} + \sum_{i=1}^{N^{(\mu,\nu)}} \sum_{j=1}^{i} w_{ij}^{(\mu,\nu)} (x_i^{(\mu,\nu)} x_j^{(\mu,\nu)}) \tag{6.2}$$

where $w_0^{(\mu,\nu)}, w_i^{(\mu,\nu)}, i = \overline{1, N^{(\mu,\nu)}}, w_{ij}^{(\mu,\nu)}, i = \overline{1, N^{(\mu,\nu)}}, j \le i$ are the weight coefficients, and they form the vector of the memory $w^{(\mu,\nu)}, u^{(\mu,\nu)}$ is the extended input vector, and $x_i^{(\mu,\nu)}$ is the i-th component of the $N^{(\mu,\nu)}$ - dimensional input vector $x^{(\mu,\nu)}$,

The combiner, or discriminant function of order k is the case in which the discriminant function is a polynomial of order k with respect to the components of the input vector. We will say that the MNN is of the k-th order if it consists of neurons of the k-th order.

The Activation Function: The activation function (a non-dynamic non-linear function) $\psi^{(\mu,\nu)}(\cdot)$ transforms the discriminant function $net^{(\mu,\nu)}(\cdot)$ to the output $y^{(\mu,\nu)}$ of the neuron. Usually all neurons have the same activation function so that $\psi^{(\mu,\nu)}(\cdot) = \psi(\cdot)$, and the (μ,ν)-neuron can be described by the following mathematical model

$$y^{(\mu,\nu)} = \psi(net^{(\mu,\nu)}). \tag{6.3}$$

There are a number of dichotomous classifications for the activation function [7]:

1. differentiable/nondifferentiable,
2. pulse-like/step-like,
3. positive/zero mean.

Differentiable functions are needed for some adaptation algorithms such as the back-propagation algorithm [6] and the recursive prediction error (RPE) algorithm [8], whereas nondifferentiable functions are needed to give a true binary output.

It is possible to create pulse-like functions from step-like functions and vice versa without differentiation via interconnection of two or more neurons.

Some standard activation functions are given in Table 6.1.

In the following part of this chapter we will discuss fully connected MNNs of the first order: the output vector of the (μ-1)-th layer is the input vector for all neurons of the μ-th layer and the MNN consists of neurons of the first order.

Table 6.1. Activation functions

Name	Formula $\psi(x)$	Characteristics
Threshold	+1 *if* $x>0$ *else* 0	non-differentiable, step-wise, non-negative
Threshold	+1 *if* $x>0$ *else* -1	non-differentiable, step-wise, zero-mean
Sigmoid	$\psi_s(x) = 1/(1+e^{-x})$	differentiable, step-wise, non-negative
Hyperbolic tangent	$\tanh(x) = 2\psi_s(2x) - 1$	differentiable, step-wise, zero-mean
Gaussian	$\psi_s(x) = e^{-x^2/\sigma^2}$	differentiable, pulse-like, non-negative

6.1.3 The Input-Output Mapping of a Multilayer Network of the First Order

For a fully connected MNN the outputs $y^{(\mu-1,v)}, v = \overline{1, N_{\mu-1}}$ of the neurons in the (μ-1)-th layer are fed to the inputs of all the neurons of the following, μ-th layer, whose output can be described by the equation:

$$y^{(\mu,v)} = \psi(w_0^{(\mu,v)} + \sum_{i=1}^{N_\mu} w_i^{(\mu,v)} y^{(\mu-1,i)}) = \psi(w^{T(\mu,v)} u^{(\mu,v)}), v = \overline{1, N_\mu}, \quad (6.4)$$

where $u^{(\mu,v)}$ is the extended input vector for the (μ, v)-neuron.

Let

$$Y^{(\mu)} = (y^{(\mu,1)}, y^{(\mu,2)}, ..., y^{(\mu,N_\mu)})^T \quad (6.5)$$

be the output vector of the μ-th layer,

$$W^{(\mu)} = (w^{(\mu,1)}, w^{(\mu,2)}, ..., w^{(\mu,N_\mu)})^T = (W_0^{(\mu)}, W_1^{(\mu)}) \tag{6.6}$$

be the $N_\mu * (N_{\mu-1} + 1)$-dimensional matrix of the memory of the neurons of the μ-th layer,

$$W_0^{(\mu)} = (w_0^{(\mu,1)}, w_0^{(\mu,2)}, ..., w_0^{(\mu,N_\mu)})^T \tag{6.7}$$

is the N_μ-dimensional vector which consists of threshold elements $w_0^{(\mu,\nu)}$. Now we can express equation (6.4) in vector-matrix form

$$Y^{(\mu)} = \Psi\left(W_0^{(\mu)} + W_1^{(\mu)} Y^{(\mu-1)}\right), \tag{6.8}$$

where Ψ is a non-linear operator which acts on each component of the vector $W_0^{(\mu)} + W_1^{(\mu)} Y^{(\mu-1)}$.

We have derived a non-linear recursive equation with respect to layer number μ. Iterating (6.8), we can write the expression for the output of the MNN:

$$Y^{(M)} = \Psi^{(M)}(W_0^{(M)} + W_1^{(M)}\Psi^{(M-1)}(W_0^{(M-1)} + W_1^{(M-1)}\Psi^{(M-2)}(...$$
$$...(W_0^{(2)} + W_1^{(2)}\Psi^{(1)}(W_0^{(1)} + W_1^{(1)}x))...))) = \Psi^{(M)}\Psi^{(M-1)}...\Psi^{(1)} x. \tag{6.9}$$

It follows from (6.9) that the output of the neural network is a complex non-linear function of the input vector x with elements of the weight matrices as parameters. This means that the MNN can be used as a non-linear approximator.

6.1.4 Approximation

A number of results have been published showing that MNNs with only two layers can approximate arbitrarily well a continuous function of n real variables. Four theorems concerning the problem of approximation will be cited here.

Theorem 6.1. (*Kolmogorov*, 1957, [9], also Kolmogorov, 1956, and Arnold, 1957) For any integer $n \geq 2$ there exist fixed continuous real (monotonically increasing) functions $\psi^{pq}(x)$ on the unit interval $I^1 = [0,1]$ such that each continuous function $f(x_1, ..., x_n)$ defined on the n-dimensional unit cube I^n can be written in the form

$$f(x_1,...,x_n) = \sum_{q=1}^{2n+1} \chi_q \left[\sum_{p=1}^{n} \psi^{pq}(x_p) \right],$$
(6.10)

where the $\chi_q(y)$ are properly chosen real and continuous functions of one variable.

Theorem 6.2. (*Sprecher*, 1964, [10]) There exist constants λ_p and fixed continuous increasing functions $\phi_q(x)$ on $I^1 = [0,1]$ such that each continuous function $f(x_1,...,x_n)$ on I^n can be written in the form

$$f(x_1,...,x_n) = \sum_{q=1}^{2n+1} g_q \left[\sum_{p=1}^{n} \lambda_p \phi_q(x_p) \right]$$
(6.11)

where g is a properly chosen continuous function of one variable.

Definition 6.1. We say that $\sigma(t)$ is sigmoidal if

$$\sigma(t) = \begin{cases} 1 & as\ t \to \infty \\ 0 & as\ t \to -\infty \end{cases}.$$

Theorem 6.3. (*Cybenko*, 1988, 1989, [11]) Let σ be any continuous sigmoidal function. Then finite sums of the form

$$G(x) = \sum_{j=1}^{N} \alpha_j \sigma(y_j^T x + \theta_j)$$
(6.12)

are dense in the space of continuous functions $C(I_n)$. In other words, given any $f \subset C(I_n)$ and $\varepsilon > 0$, there is a sum, $G(x)$, of the above form, for which $|G(x) - f(x)| < \varepsilon$ for all $x \in I_n$.

Theorem 6.4. (*Funahashi*, 1989, [11]) Let $\phi(x)$ be a non-constant, bounded and monotonically increasing continuous function. Let K be a compact subset of R^n and $f(x_1,...,x_n)$ be a real valued continuous function on K. Then for any $\varepsilon > 0$, there exist an integer N and real constants $c_i, \sigma_i\ (i = 1,...,N)$, $w_{ij}\ (i = 1,...,N, j = 1,...,n)$ such that

$$\tilde{f}(x_1,...,x_n) = \sum_{i=1}^{N} c_i \phi \left(\sum_{j=1}^{n} w_{ij} x_i - \sigma_i \right)$$
(6.13)

satisfies $\max_{x \in K} \left| f(x_1,...,x_n) - \tilde{f}(x_1,...,x_n) \right| < \varepsilon$, where the σ_i in (6.11) are bias weights.

These theorems show that arbitrary continuous functions of n variables can be approximated arbitrarily well by continuous neural networks with only two layers and any sigmoidal nonlinearity.

In the first two theorems the number k of neurons necessary for exact approximation of any non-linear function of n real variables is given: $k=2n+1$.(Kolmogorov, Sprecher).

In Theorem 6.2 the non-linear functions $\psi^{pq}(x_p)$ of Theorem 6.1 are replaced by $\lambda_p \phi_q(x_p)$, and the functions $\chi_q(\cdot)$ and $g_q(\cdot)$ depend on the approximated function.

Theorems 6.3 and 6.4 are similar to each other. It should be noted that in Theorem 6.4 the activation functions are taken to be monotonically increasing whereas no monotonicity is required for the activation functions in Theorem 6.3.

Unfortunately, these Theorems do not state how many hidden layers should be used and how many neurons are needed in every layer.

6.2 Learning Algorithms for Single Elements of Multilayer Neural Networks

The learning process of multilayer neural networks is carried out by comparison of the network's outputs (i.e., the outputs of the neurons of the last layer) with the corresponding teacher's instructions. Information about the ideal outputs of the neurons of hidden layers is absent. Despite this, knowledge of the structure of a MNN allows us to calculate the correcting signals not only for the neurons of the last layer but also for the neurons of the other layers.

The neurons of the last layer are in a unique position, the learning process of these neurons is the same as for a separate neuron because in this case the teacher's instructions are available.

A neuron is described by the following equation (6.3)

$$y = \psi(w^T u),$$
(6.14)

where u is an extended input vector of the neuron and w is the vector of neuron's memory. The closeness of the neuron's output $y(n)$ to the teacher's instruction y^* (see Chapter 1) is described by a performance index $J(w)$. For simplicity, we will restrict the analysis to the following performance index:
if the total number P of input vectors is fixed then

$$
\begin{aligned}
J(w) &= \frac{1}{P} \sum_{m=1}^{P} Q(y^*(m), y(m)) = \frac{1}{P} \sum_{m=1}^{P} (y^*(m) - \psi(w^T u(m))^2 \\
&= \frac{1}{P} \sum_{m=1}^{P} \varepsilon(w;m)^2
\end{aligned}
\tag{6.15}
$$

whereas

$$
\begin{aligned}
J(w;n) &= \frac{1}{n} \sum_{m=1}^{n} Q(y^*(m), y(m)) = \frac{1}{n} \sum_{m=1}^{n} (y^*(m) - \psi(w^T u(m))^2 \\
&= \frac{1}{n} \sum_{m=1}^{n} \varepsilon(w;m)^2
\end{aligned}
\tag{6.16}
$$

if the input vector $u(m)$ is random. In (6.15) and (6.16) m is the number of the current input vector. The goal of the neuron is to find the minimum of the performance index and it can be accomplished by changing the neuron's memory employing learning algorithms, as discussed in Chapter 3.

6.2.1 Differentiable Activation Functions

If $J(w)$ is differentiable then performance indexes (6.15) and (6.16) are differentiable and we can use the gradient algorithm or Newton's algorithm for the learning process. The instantaneous gradient and Hessian corresponding to the performance index (6.15) and (6.16) can be written in the form

$$
\nabla_w Q(\varepsilon(w;m) = -2\varepsilon(w;m)\psi'(w^T u(m))u(m),
\tag{6.17}
$$

$$
\begin{aligned}
\nabla_w^2 Q(\varepsilon(w;m) &= \\
2\big[(\psi'(w^T u(m)))^2 &- \varepsilon(w;m)\psi''(w^T u(m))\big]u(m)u(m)^T \\
&= 2q(w;m)u(m)u^T(m)
\end{aligned}
\tag{6.18}
$$

respectively, where the scalar multiplier

$$q(w;m) = \left[(\psi'(w^T u(m)))^2 - \varepsilon(w;m)\psi''(w^T u(m))\right] \tag{6.19}$$

is a function of the memory w.

The Deterministic Case: Expressions (6.17) and (6.18) allow us to write the learning algorithm corresponding to function (6.15) (see Chapter 3).

The Gradient Algorithm:

$$w(n) = w(n-1) + \gamma \sum_{m=1}^{P} \varepsilon(w(n-1);m)\psi'(w^T(m-1)u(m))u(m), \tag{6.20}$$

where γ is the step size and $w(0) = w_0$ is the initial value of the memory.

Newton's Algorithm:

$$w(n) = w(n-1) +$$
$$\Gamma(w(n-1))\sum_{m=1}^{P} \varepsilon(w(n-1);m)\psi'(w^T(m-1)u(m))u(m)' \tag{6.21}$$

where the matrix

$$\Gamma(w(n-1)) = \left(\sum_{m=1}^{P} q(w(n-1);m)u(m)u^T(m)\right)^{-1} \tag{6.22}$$

can be calculated using the matrix inversion lemma (see Chapter 4).

The Stochastic Case: Algorithms corresponding to function (6.16) have the form:

The Gradient Algorithm:

$$w(n) = w(n-1) + \gamma(n)\varepsilon(w(n-1);n)\psi'(w^T(m-1)u(n))u(n), \tag{6.23}$$

where $\gamma(n)$ is the step size (usually $\gamma(n) \to 0$ *if* $n \to \infty$), and $w(0) = w_0$ is the initial value of the memory.

Newton's Algorithm:

$$w(n) = w(n-1) + \Gamma(w(n-1))\varepsilon(w(n-1);m)\psi'(w^T(m-1)u(m))u(m), \quad (6.24)$$

where matrix

$$\Gamma(w(n-1)) = \left(\sum_{m=1}^{P} q(w(m-1);m)u(m)u^T(m)\right)^{-1} \quad (6.25)$$

can be calculated recursively.

6.2.2 Non-Differentiable Activation Functions

Suppose that the activation function is a symmetrical threshold element and the teacher's instructions are binary. In this case the neuron is used to produce a binary output, y=(+1,-1) and is called an α-Perceptron (Rosenblatt, 1960), or Adaline (adaptive linear element), (Widrow, Hoff). The details concerning these algorithms can be found in [13].

The learning process for the α-Perceptron can be written in the form (Rosenblatt, [13]):

$$w(n) = w(n-1) + \alpha \frac{(y^*(n) - sign(w^T(n-1)u(n))}{2u^T(n)u(n)} u(n), \quad (6.26)$$

where α is normally equal to one;

or in the following form (Mays, [13]

$$w(n) = \begin{cases} w(n-1) + \alpha\, \varepsilon(w(n-1);n)\dfrac{u(n)}{2u^T(n)u(n)} & \text{if } abs(w^T(n-1)) \geq \gamma \\ w(n-1) + \alpha\, y^*(n)\dfrac{u(n)}{u^T(n)u(n)} & \text{if } abs(w^T(n-1)) < \gamma \end{cases} \quad (6.27)$$

If the dead zone parameter $\gamma = 0$, then Mays' increment adaptation algorithm reduces to Rosenblatt's algorithm.

It is possible to explain algorithms (6.26) and (6.27) as projection procedures in the parametric space **W**.

In the following section we will derive a learning algorithm for the neuron with a hard-limited characteristic directly based on the projection procedure.

Let us consider the learning process of a two-input Adaline element

$$y(n) = sign(w_0 + w_1 x_1(n) + w_2 x_2(n)),$$ (6.28)

where $sign(z) = 1$ *if* $z > 0$ *otherwise* $sign(z) = 0$, with four binary inputs $x(n)$ and the corresponding teacher's instructions y^*:

n	x_1	x_2	y^*	
1	+1	+1	+1	
2	−1	+1	+1	(6.29)
3	−1	−1	+1	
4	+1	−1	−1	

Figure 6.2(a) represents all four possible binary inputs on the plane X (large dots A, B, C, and D) to this element with corresponding teacher's instructions +1 or -1. It follows from this figure that in this case there exists a line $L(w)$: $w_0 + w_1 x_1 + w_2 x_2 = 0$ which separates the inputs patterns $x(n)$, $n = 1, 2, 3, 4$ into two categories according to the teacher's instructions (the input space is linearly separable).

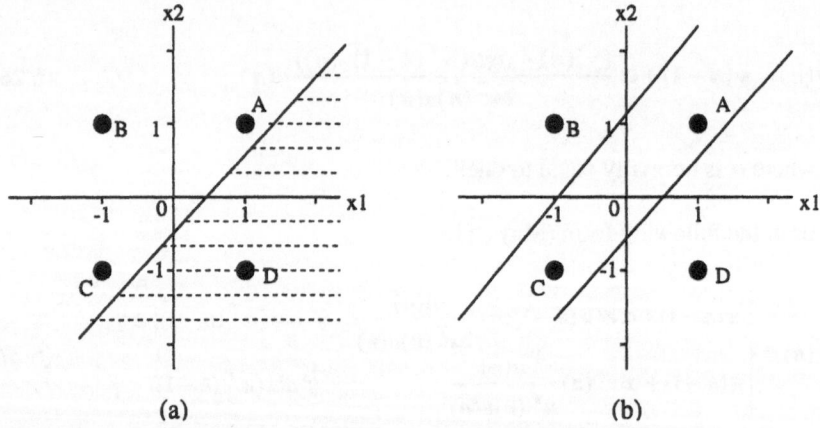

(a)   (b)

Fig 6.2. Examples of linearly and non-linearly separable problems.

The task of finding the separation line $L(w)$ is equivalent to estimation of the coefficients
w_0, w_1 and w_2.

We can write the following linear system of homogeneous inequalities corresponding to (6.28) and (6.29)

$$w_0 + w_1 + w_2 > 0,$$
$$w_0 - w_1 + w_2 > 0,$$
$$w_0 - w_1 - w_2 > 0,$$
$$w_0 + w_1 - w_2 < 0.$$

(6.30)

Obviously, if vector $w^* = (w_0^*, w_1^*, w_2^*)^T$ is a solution of the system (6.30),

then so is vector $\overline{w}^* = \left(\dfrac{w_0^*}{|w_0^*|}, \dfrac{w_1^*}{|w_0^*|}, \dfrac{w_2^*}{|w_0^*|} \right)^T$. This allows us to normalise the

solution so that we can seek the solution of one of the two following systems of linear inequalities:

$$1 + w_1 + w_2 > 0,$$
$$1 - w_1 + w_2 > 0,$$
$$1 - w_1 - w_2 > 0,$$
$$1 + w_1 - w_2 < 0.$$

(6.31)

$$-1 + w_1 + w_2 > 0,$$
$$-1 - w_1 + w_2 > 0,$$
$$-1 - w_1 - w_2 > 0,$$
$$-1 + w_1 - w_2 < 0.$$

(6.32)

It follows from (6.31) and (6.32) that in the separable case the threshold w_0 can be set to be equal to +1 or -1.

The lines $L_1(w) = 0, ..., L_4(w) = 0$ in the plane (w_1, w_2) are given in the figures 6.3(a) and Fig. 6.3(b) for $w_0 = +1$ and $w_0 = -1$ together with corresponding shaded region of solutions: the numbers of correct outputs of the neuron in the different regions are given in the circles. It follows from the comparison of Fig. 6.3(a) and Fig. 6.3(b) that only $w_0 = +1$ can be used as a solution and the solution w^* is any point in the set S.

To construct a learning algorithm corresponding to the task (6.29) we will use the projection procedure:

1. Let $w_0 = 1$,

2. The point $w(o) = (w_1(0), w_2(0))^T$ is arbitrary assigned,

3. If the point $w(0)$ is a solution of the first inequality then this point remains unchanged, in the opposite case this point is orthogonally projected onto the first line $L_1(w) = 0$ with a small overstepping of the line (the parameter γ: $1 < \gamma < 2$) and this projection is equal to $w(1)$.
4. The construction principle described above is applied to all successive estimates $w(n)$.

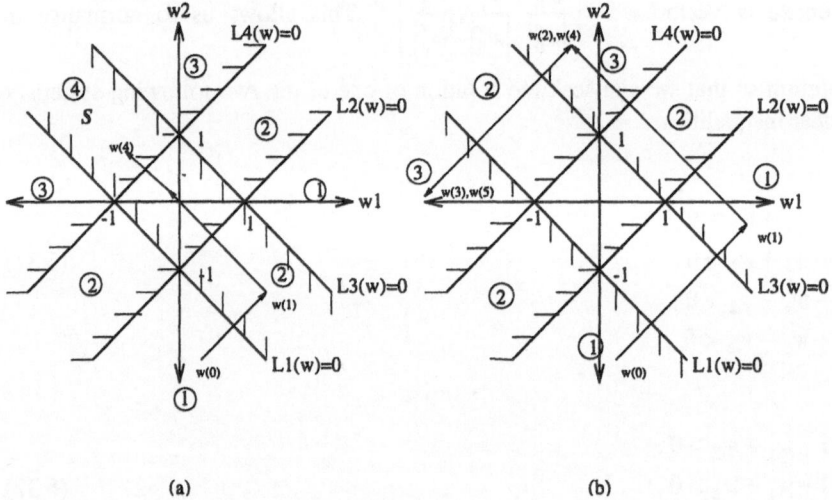

(a)   (b)

Fig 6.3. Weight space representation of Eq. (6.31,6.32)

The algorithm will converge in a finite number of steps if S exists and is non-degenerate. For arbitrary dimension N of the input vector x the algorithm described above consists of two procedures for

$$w_0 = 1 \text{ if } \delta = -1 \text{ and } w_0 = -1 \text{ if } \delta = +1: \tag{6.33}$$

$$w(n) = \begin{cases} w(n-1) + \gamma \dfrac{\delta - w^T(n-1)x(n)}{x(n)^T x(n)} x(n) & \text{If the neuron's output doesn't coincide with the teacher's instruction} \\ \\ w(n-1) & \text{Otherwise} \end{cases} \tag{6.34}$$

If a solution of the system of linear inequalities exists then algorithm (6.34) will converge to the region of solution, and the value w_0 will be defined from condition (6.33).

The sequences of points $w(n)$ calculated according to algorithm (6.33) and (6.34) are given in figures 6.3(a) and figure 6.3(b). The parameter γ is equal to 1.5. If $w_0 = +1$ then convergence is accomplished in 4 steps (Fig. 3.a). If $w_0 = -1$ we have an asymptotically oscillating regime between regions for which the number of correct outputs is at a maximum and equal to three.

An example of non-linearly separable inputs is given in Fig. 6.2(b) (the simplest variant of the parity problem). The four binary inputs $x(n)$ and the corresponding teacher's instructions y^* are:

n	x_1	x_2	y^*
1	+1	+1	+1
2	−1	+1	−1
3	−1	−1	+1
4	+1	−1	−1

The corresponding inequalities are for $w_0 = 1$ and $w_0 = -1$, are respectively

$$1 + w_1 + w_2 > 0 ,$$
$$1 - w_1 + w_2 < 0 , \tag{6.35}$$
$$1 - w_1 - w_2 > 0 ,$$
$$1 + w_1 - w_2 < 0 .$$

$$-1 + w_1 + w_2 > 0 ,$$
$$-1 - w_1 + w_2 < 0 , \tag{6.36}$$
$$-1 - w_1 - w_2 > 0 ,$$
$$-1 + w_1 - w_2 < 0 .$$

From figures 6.4(a) and 6.4(b) it follows that the neuron cannot solve this problem.

It is easily shown that a two layer neural network with two neurons in the first layer (one from the region A and the other from region B, Fig. 6.4(a)) and one neuron (from the region C, Fig. 6.4(b)) in the second layer can solve the parity problem.

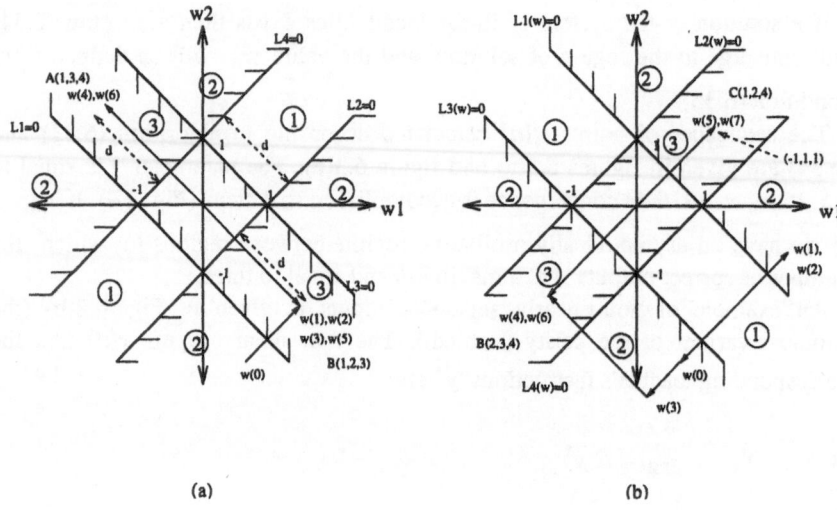

(a)   (b)

Fig 6.4. Weight space representation of Eq. (6.35,6.36)

References

[1] Y. H. Pao, Adaptive pattern recognition and neural networks. Addison-Wesley , 1989.

[2] S. Chen, S. A.Billings, and P. M. Grant, "Non-linear system identification using neural networks, " Int. J. Contr., vol. 51, No. 6, pp. 1191-1214, 1990.

[3] W. T. Miller, R. S. Sutton, and P. J. Werbos, Eds., Neural networks for control. The MIT Press, 1990.

[4] K. Warwick, G. W. Irwin, and K. J. Hunt, Eds., Neural networks for control and systems. London: Peter Peregrinus, 1988.

[5] B. Horne, M. Jamshidi, and N.Vadice, "Neural networks in robotics: a survey", J. of Intelligent and Robotics Systems, No. 3, pp. 51-66, 1990.

[6] D. E. Rumelhart, G. E. Hinton, and R. J. Williams, "Learning internal representations by error propagation", Parallel Distributed Processing, V.1, ch. 8, pp. 675-695, D. E. Rumelhart and J. L. McCelland, Eds., Cambridge, MA: MIT Press, 1986.

[7] K. J. Hunt, D. Sbarbaro, R. Zbikowski and P. J. Gawthrop, "Neural networks for control systems-a survey", Automatica, vol. 28, No. 6, pp. 1083-1112, 1992.

[8] S. A. Billings, H. B. Jamaluddin and S. Chen, "Properties of neural networks with applications to modeling non-linear dynamical systems", Int. J. Control, vol. 55, No. 1, pp. 193-224, 1992.

[9] A. N. Kolmogorov, "On the representation of continuous functions of several variables by superpositions of continuous functions of one variable and addition", Doklady Akademii Nauk SSSR, vol. 114, No.5, pp. 953-956.

[10] D. A. Sprecher, "On the structure of continuous functions of several variables, "Transactions of the American Mathematical Society", 115, 340-355.

[11] G. Cybenko, "Approximation by superpositions of a sigmoidal function, "Mathematics of Control, Signals, and Systems", 2, pp. 303-313, 1989.

[12] K. Funahashi, " On the approximate realization of continuous mappings by neural networks", Neural Networks, 2, pp. 183-192, 1989.

[13] B. Widrow, M. A. Lehr, "30 Years of adaptive neural networks: Perceptron, Madaline, and Backpropagation", Proceedings of IEEE, vol. 78, No. 9, pp. 1415-1442.

7 Learning Algorithms for Neural Networks

7.1 The Back-Propagation Algorithm for MNN Learning

7.1.1 Main Equations

The optimal value w for the memory of a multilayer neural network is determined by the minimisation of the performance index $J(w)$ which characterises the proximity of the network outputs to the teacher's instructions. Here w is a composite vector which consists of the memory vectors $w^{(\mu,\nu)}$ of the individual neurons.

The learning process of multilayer neural networks is carried out by comparison of the network's outputs (i.e., the outputs of the neurons of the last layer) with the corresponding teacher's instructions. Information about the ideal outputs of the neurons of hidden layers is absent. Despite this, the knowledge of the structure of a MNN allows us to calculate the correcting signals not only for the neurons of the last layer but also for the neurons of the other layers.

We shall use the following notation from the previous chapter:

Let:

the ν-th neuron of the μ-th layer be the (μ,ν)-neuron,

$y^{(\mu,\nu)} = \psi(net^{(\mu,\nu)})$, $\nu = \overline{1, N_\mu}$, $\mu = \overline{1, M}$ be the output of the (μ,ν)-neuron,

$net^{(\mu,\nu)} = w^{T(\mu,\nu)}u^{(\mu,\nu)}$, $\nu = \overline{1, N_\mu}, \mu = \overline{1, M}$ be the output of the linear combiner,

$\psi(\cdot)$ be a sigmoidal activation function,

$w^{(\mu,\nu)}, u^{(\mu,\nu)}$ be the memory vector and the extended input vector, respectively,

$Y^{(\mu)} = (y^{(\mu,1)}, y^{(\mu,2)}, ..., y^{(\mu,N_\mu)})^T$ be the output vector of the μ-th layer,

$Y^{(M)} = (y^{(M,1)}, y^{(M,2)}, ..., y^{(M,N_M)})^T$ be the output vector of the last layer,

$Y^{(*)} = (y^{(*,1)}, y^{(*,2)}, ..., y^{(*,N_M)})^T$ be the vector of the teacher's instructions,

$W^{(\mu)} = (w^{(\mu,1)}, w^{(\mu,2)}, ..., w^{(\mu,N_\mu)})^T = (W_0^{(\mu)}, W_I^{(\mu)})$ be the $N_\mu * (N_{\mu-1} + 1)$ -

dimensional memory matrix which consists of the memory vectors $w^{(\mu,\nu)}$ of the neurons of the μ-th layer,

$W_0^{(\mu)} = (w_0^{(\mu,1)}, w_0^{(\mu,2)}, ..., w_0^{(\mu,N_\mu)})^T$ be the N_μ -dimensional vector which consists of the threshold elements $w_0^{(\mu,\nu)}$.

The closeness of the MNN's output $Y^{(M)}(m)$ to the teacher's instruction $Y^{(*)}(m)$ (see Chapter 1) is described by an instant performance index $Q(\varepsilon(w;m))$ where

$$\varepsilon(w;m) = Y^{(*)}(m) - Y^{(M)}(m) \tag{7.1}$$

is the error vector of the MNN and the index m is the discrete time. The instant performance index $Q(\varepsilon(w;m))$ can be, for example, of the form

$$Q(\varepsilon(w;m)) = \frac{1}{2}\varepsilon^T(w;m)\varepsilon(w;m)$$

or

$$Q(\varepsilon(w;m)) = \frac{1}{2}\varepsilon^T(w;m)R\,\varepsilon(w;m), \tag{7.2}$$

where R is a positive definite matrix.

The performance index for all possible input vectors is:

$$J(w) = \frac{1}{P}\sum_{m=1}^{P} Q(\varepsilon(w;m)) \tag{7.3}$$

if the total number P of the input vectors is fixed, or

$$J(w;n) = \frac{1}{n} \sum_{m=1}^{n} Q(\varepsilon(w;m)) \tag{7.4}$$

if the input vector is random (see Chapter I).

The search for the optimal memory vector w of the MNN is fulfilled with the gradient method for minimisation with respect to function (7.3) or (7.4), for example [1]:

$$w(n) = w(n-1) - \gamma_1 \, \nabla_w J(w(n-1);n), \tag{7.5}$$

where $\nabla_w J(w(n-1);n)$ is the gradient of the performance index $J(w)$ at point $w(n-1)$ and $\gamma_1 > 0$ is a constant characterising the length of the correction step.

From previous chapters it follows that if the performance index can be described by Eqs.(7.3) and (7.4) then the algorithm (7.5) takes the form:

$$w(n) = w(n-1) - \gamma_2 \sum_{m=1}^{P} \nabla_w Q(\varepsilon(w(n-1);m)) \tag{7.6}$$

if the total number P of the input vectors is fixed, and

$$w(n) = w(n-1) - \gamma \, \nabla_w Q(w(n-1);n) \tag{7.7}$$

if the input vector is random.

The values of γ_2 and γ are usually assumed to be constant in the procedures (7.6) and (7.7), and as a result the procedure (7.7) only provides convergence to the neighbourhood of the optimal value for the memory w. In order to find the optimal value of w that minimises function (7.4), the coefficient γ should be a function of n as in probabilistic algorithms of optimisation and adaptation [2] (see Chapter 3) .

The main difficulty in implementing procedure (7.6) or (7.7) is caused by the complex dependence of the network output on the neuron's memory and in the calculation of the gradient $\nabla_w J(w(n-1);n)$ or $\nabla_w Q(w(n-1);n)$.

An efficient method for the gradient calculation based on the successive nature of the transformation of the input signal to an MNN is realised by the back-propagation method [3].

Let us write the expression for the gradient ∇Q with respect only to memory vector $w^{(\mu,\nu)}$ omitting all the non-necessary indexes for clarity:

$$\nabla_{w^{(\mu,\nu)}}Q = \frac{\partial Q}{\partial y^{(\mu,\nu)}}\nabla_{w^{(\mu,\nu)}}y^{(\mu,\nu)} = \frac{\partial Q}{\partial y^{(\mu,\nu)}}\frac{\partial y^{(\mu,\nu)}}{\partial net^{(\mu,\nu)}}\nabla_{w^{(\mu,\nu)}}net^{(\mu,\nu)}. \qquad (7.8)$$

The expressions for the second and the third factors are, obviously:

$$\frac{\partial y^{(\mu,\nu)}}{\partial net^{(\mu,\nu)}} = \psi'(net^{(\mu,\nu)}) \qquad (7.9)$$

and

$$\nabla_{w^{(\mu,\nu)}}net^{(\mu,\nu)} = u^{(\mu,\nu)} \qquad (7.10)$$

respectively.

The main problem is to calculate the first factor $s^{(\mu,\nu)} = \dfrac{\partial Q}{\partial y^{(\mu,\nu)}}$. For this partial derivative a recursive equation takes place in the backwards direction in decreasing numbers of the neural network layers. For deriving this equation let us remember that the outputs $y^{(\mu+1,j)}, j = \overline{1, N_{\mu+1}}$ of the neurons of the $(\mu+1)$-th layer are functions of the output $y^{(\mu,\nu)}$ by virtue of the structure of the network.

7.1.2 Fully Connected MNNs

For fully connected MNNs the output of the (i,ν)-neuron is an input for all neurons of the following $(i+1)$-th layer. Using the differentiation chain rule we can write the following recursive equations with respect to variable s:

$$s^{(i,\nu)} = \frac{\partial Q}{\partial y^{(i,\nu)}} = \sum_{j=1}^{N_{i+1}}\frac{\partial Q}{\partial y^{(i+1,j)}}\frac{\partial y^{(i+1,j)}}{\partial y^{(i,\nu)}} = \sum_{j=1}^{N_{i+1}}s^{(i+1,j)}d_{\nu,j}^{(i+1)}, \qquad (7.11)$$

$$v = \overline{1, N_i}$$

If we introduce the sensitivity vector $S^{(i)}$ of the instant function Q with respect to vector $Y^{(i)}$:

$$S^{(i)} = \nabla_{Y^{(i)}} Q = \left(\frac{\partial Q}{\partial y^{(i,1)}}, \frac{\partial Q}{\partial y^{(i,2)}}, ..., \frac{\partial Q}{\partial y^{(i,N_i)}} \right)^T \tag{7.12}$$

$$= (s^{(i,1)}, s^{(i,2)}, ..., s^{(i,N_i)})^T$$

and the $N_i \cdot N_{i+1}$ -dimensional matrix $D^{(i+1)}$ of the transient sensitivity of the vector $Y^{(i+1)T}$ with respect to vector $Y^{(i)}$:

$$D^{(i+1)} = \left(\frac{\partial y^{(i+1,j)}}{\partial y^{(i,v)}} \right) = (d_{v,j}^{(i+1)}) = \nabla_{Y^{(i)}} Y^{(i+1)T} \tag{7.13}$$

then we can write the equations (7.11) in the recursive vector-matrix form:

$$S^{(i)} = D^{(i+1)} S^{(i+1)}, \quad i = \overline{M-1,1}. \tag{7.14}$$

Equation (7.14) allows one to calculate $S^{(i)}$ recursively starting from the last, M-th layer for which the calculation of the sensitivity vector is very simple. For example, if the instant performance index Q has the form $Q = \frac{1}{2} \varepsilon^T \varepsilon$ or $Q = \frac{1}{2} \varepsilon^T R \varepsilon$ where R is a positive definite matrix and $\varepsilon = Y^{(*)} - Y^{(M)}$ is the error vector, then the sensitivity vector is $S^{(M)} = -\varepsilon$ or $S^{(M)} = -R\varepsilon$, respectively. Since the sensitivity vector $S^{(i)}$ may be considered as the generalised error of the network and the iterations are carried out in the backwards direction, then the method is called the error back-propagation method.

From expression (7.8) it follows that the gradient of Q with respect to the matrix memory $W^{(\mu)}$ of the μ-th layer for fully connected MNN ($u^{(\mu,v)} \equiv u^{(\mu)}$) can be written in the form

$$\nabla_{W^{(\mu)}}Q \ =$$

$$
\begin{pmatrix}
\dfrac{\partial Q}{\partial y^{(\mu,1)}}\psi'(net^{(\mu,1)})u^{(\mu)}, \dfrac{\partial Q}{\partial y^{(\mu,2)}}\psi'(net^{(\mu,2)})u^{(\mu)}, \dots, \\[4mm]
\dfrac{\partial Q}{\partial y^{(\mu,N_\mu)}}\psi'(net^{(\mu,N_\mu)})u^{(\mu)}
\end{pmatrix} \ =
$$

$$
u^{(\mu)}\begin{pmatrix}
\dfrac{\partial Q}{\partial y^{(\mu,1)}}\psi'(net^{(\mu,1)}), \dfrac{\partial Q}{\partial y^{(\mu,2)}}\psi'(net^{(\mu,2)}), \dots, \\[4mm]
\dfrac{\partial Q}{\partial y^{(\mu,N_\mu)}}\psi'(net^{(\mu,N_\mu)})
\end{pmatrix} \tag{7.15}
$$

Let us introduce the $N_\mu \cdot N_\mu$ -diagonal matrix with the diagonal elements $\psi'(net^{(\mu,k)})$, $k = \overline{1, N_\mu}$:

$$Diag(\psi'(net^{(\mu)})). \tag{7.16}$$

Using expressions (7.12) for the sensitivity vector $S^{(i)}$ for $i = \mu$ and (7.16) for the matrix $Diag(\psi'(net^{(\mu)}))$, we can write $\nabla_{W^{(\mu)}}Q$ (7.15) in a more convenient form

$$\nabla_{W^{(\mu)}}Q \ = u^{(\mu)}Diag(\psi'(net^{(\mu)}))S^{(\mu)T} . \tag{7.17}$$

Equations (7.14) and (7.17) are the main expressions for the back-propagation algorithm and they are valid for fully connected MNNs of any order. The order of the MNN affects only matrix $D^{(i+1)}$ (7.13) in the recursive equation (7.14).

7.1.3 The Transient Sensitivity Matrix $D^{(i+1)}$ for Fully Connected MNNs of the First Order

The recursive equation which describes the connection between outputs of the i-th and $(i+1)$-th layers of fully connected MNN of the *first* order has the form (see Chapter 6):

$$Y^{(i+1)} = \Psi^{(i+1)}\big(W_0^{(i+1)} + W_I^{(i+1)}Y^{(i)}\big), \tag{7.18}$$

where $\Psi^{(i+1)}(a) \equiv (\psi(a_1), \psi(a_2), ..., \psi(a_{N_{i+1}}))^T$ is a non-linear vector-operator and a is any N_{i+1}-dimensional vector. In equation (7.18) matrices $W_0^{(i+1)}$, $W_I^{(i+1)}$ are given by expression

$$W^{(i+1)} = (w^{(i+1,1)}, w^{(i+1,2)}, ..., w^{(i+1,N_\mu)})^T = (W_0^{(i+1)}, W_I^{(i+1)}) \tag{7.19}$$

where $W^{(i+1)}$ is the $N_{i+1} \cdot (N_i + 1)$-dimensional matrix of the memory of the neurons of the $(i+1)$-th layer and

$$W_0^{(i+1)} = (w_0^{(i+1,1)}, w_0^{(i+1,2)}, ..., w_0^{(i+1,N_\mu)})^T \tag{7.20}$$

is the N_{i+1}-dimensional vector which consists of the threshold elements $w_0^{(i+1,v)}$ of each neuron and $W_1^{(i+1)}$ is $N_{i+1} \cdot N_i$-matrix.

The $N_i \cdot N_{i+1}$-matrix

$$D^{(i+1)} = \left(\frac{\partial y^{(i+1,j)}}{\partial y^{(i,v)}} \right) = (d_{v,j}^{(i+1)}) = \nabla_{Y^{(i)}} Y^{(i+1)T}$$

of the equation (7.14) in accordance with (7.18) has the form

$$\dot{D}^{(i+1)} = (d_{v,j}^{(i+1)}) = \nabla_{Y^{(i)}} Y^{(i+1)T} = W_1^{(i+1)T} diag(\psi'(net^{(i+1)})) \tag{7.21}$$

where the diagonal elements of the $N_{i+1} \cdot N_{i+1}$-diagonal matrix $diag(\psi'(net^{(i+1)}))$ are $\psi'(net^{(i+1,1)}), \psi'(net^{(i+1,2)}), ..., \psi'(net^{(i+1,N_{i+1})})$ and the matrix $W_1^{(i+1)T}$ is defined by Eq.(7.19).

7.1.4 The Transient Sensitivity Matrix $D^{(i+1)}$ for Fully Connected MNNs of the Second Order

For fully connected MNNs the outputs $y^{(i, j)}, j = \overline{1, N_i}$ of the neurons of the i-th layer are fed to the inputs of all the neurons of the following $(i+1)$-th layer, which v-th outputs can be described by the equation:

$$y^{(i+1,v)} = \psi(w_0^{(i+1,v)} + \sum_{j=1}^{N_i} w_j^{(i+1,v)} y^{(i, j)} + \sum_{k=1}^{N_i} \sum_{j=k}^{N_i} w_{j,k}^{(i+1,v)} y^{(i, j)} y^{(i, k)}) =$$

$$= \psi \left(w_0^{(i+1,v)} + w_I^{(i+1,v)T} Y^{(i)} + Y^{(i)T} w_{II}^{(i+1,v)} Y^{(i)} \right) = \psi \left(w^{(i+1,v)T} u^{(i+1)} \right) \qquad (7.22)$$

where

$$u^{(i+1)} = (1, y_1^{(i)}, y_2^{(i)}, \ldots, y_{N_i}^{(i)}, y_1^{(i)2}, y_2^{(i)} y_1^{(i)}, y_3^{(i)} y_1^{(i)}, \ldots, y_2^{(i)2}, y_3^{(i)} y_2^{(i)}, \ldots, y_{N_i}^{(i)2})^T$$

is an extended input vector of the neurons of the $(i+1)$-th layer,

$$w^{(i+1,v)} = (w_0^{(i+1,v)}, w_1^{(i+1,v)}, \ldots, w_{N_i}^{(i+1,v)},$$

$$w_{1,1}^{(i+1,v)}, w_{2,1}^{(i+1,v)}, \ldots, w_{N_i,1}^{(i+1,v)}, w_{2,2}^{(i+1,v)}, \ldots, w_{N_i,N_i}^{(i+1,v)})$$

is an extended vector of the memory of the $(i+1,v)$-neuron, and

$$w_{II}^{(i+1,v)} \qquad (7.23)$$

is a $N_i \cdot N_i$ -lower triangular matrix of the $(i+1,v)$ – neuron in accordance with the form of the extended input vector $u^{(i+1)}$ and extended vector of the memory.

The recursive equation which describes the connection between outputs of the i-th and $(i+1)$-th layers of fully connected MNNs of the *second* order can be now written in the following form:

$$Y^{(i+1)} = \Psi^{(i+1)} \left(W_0^{(i+1)} + W_I^{(i+1)} Y^{(i)} + \tilde{Y}^{(i)} W_{II}^{(i+1)} Y^{(i)} \right), \qquad (7.24)$$

where the quadratic form $\tilde{Y}^{(i)} W_{II}^{(i+1)} Y^{(i)}$ is included into the net-function. In Eq.(7.24) the $(N_{i+1} \cdot N_i) \cdot N_i$ – matrix $W_{II}^{(i+1)}$ and the $N_i \cdot (N_{i+1} \cdot N_i)$ – matrix $\tilde{Y}^{(i)}$ have the following block structure, respectively:

$$W_{II}^{(i+1)T} = \left(w_{II}^{(i+1,1)T}, w_{II}^{(i+1,2)T}, \ldots, w_{II}^{(i+1,N_{i+1})T} \right), \qquad (7.25)$$

$$
\tilde{Y}^{(i)} = \begin{pmatrix} Y^{(i)T} & 0 & . & 0 \\ 0 & Y^{(i)T} & . & 0 \\ 0 & 0 & . & 0 \\ 0 & 0 & . & Y^{(i)T} \end{pmatrix}.
\tag{7.26}
$$

The $N_i \cdot N_{i+1}$ -matrix of the transient sensitivity

$$
D^{(i+1)} = \left(\frac{\partial y^{(i+1,j)}}{\partial y^{(i,v)}} \right) = (d_{v,j}^{(i+1)}) = \nabla_{Y^{(i)}} Y^{(i+1)T}
$$

of equation (7.14) in accordance with Eqs.(7.22) and (7.24) has the form

$$
D^{(i+1)} = (d_{v,j}^{(i+1)}) = \nabla_{Y^{(i)}} Y^{(i+1)T} =
$$
$$
\left(W_I^{(i+1)T} + \left(W_{II}^{(i+1)T} + \tilde{W}^{(i+1)T}_{II} \right) \tilde{Y}^{(i)T} \right) diag(\psi'(net^{(i+1)})
\tag{7.27}
$$

where the matrix $\tilde{W}^{(i+1)T}_{II}$ has the same structure as matrix $W_{II}^{(i+1)T}$ but the matrices $w_{II}^{(i+1,v)}, v = \overline{1, N_{i+1}}$ are transposed:

$$
\dot{W}_{II}^{(i+1)T} = \left(w_{II}^{(i+1,1)T}, w_{II}^{(i+1,2)T}, ..., w_{II}^{(i+1,N_{i+1})T} \right),
\tag{7.28}
$$

$$
\tilde{W}^{(i+1)T}_{II} = \left(w_{II}^{(i+1,1)}, w_{II}^{(i+1,2)}, ..., w_{II}^{(i+1,N_{i+1})} \right).
\tag{7.29}
$$

The matrix $\tilde{Y}^{(i)T}$ is defined in accordance with expression (7.26).

7.1.5 Non-Fully Connected MNNs

For non-fully connected MNNs the output of the (i,v) -neuron is an input only for some of the neurons of the following $(i+1)$ -th layer. Using the differentiation chain rule we can write the following recursive equations with respect to variable s as in Eq.(7.11) :

$$s^{(i,v)} = \frac{\partial Q}{\partial y^{(i,v)}} = \sum_{j_v} \frac{\partial Q}{\partial y^{(i+1,j_v)}} \frac{\partial y^{(i+1,j_v)}}{\partial y^{(i,v)}} = \sum_{j_v} s^{(i+1,\ j_v)} d_{v,j_v}^{(i+1)} , v = \overline{1, N_i} .$$

$$(7.30)$$

Summation in expression (7.30) is performed only for neurons of the $(i+1)$-th layer which are connected to the (i,v) -neuron.

It follows from Eq.(7.30) that the back-propagation algorithm for partially connected MNNs is similar to that of the full connection case, the only difference being that some elements of matrix $D^{(i+1)}$ (7.13) will be equal to zero.

7.2 Autonomous Algorithms for Adjusting MNNs

7.2.1 Introduction

At present the back-propagation algorithm is widely used as an algorithm for adjusting the weights of multilayer neural networks. This algorithm is characterised by a recursive computation of the gradients of complex non-linear functions. Nevertheless, it follows from a number of papers (see, e.g., [4],[5]), that the back-propagation method often displays slow convergence and is subject to fall into local extrema of the function [5],[6]. However, the structure of this algorithm is not in agreement with neurobiological concepts because biological neurons cannot interchange as large an amount of information about their states as required for artificial neurons when implementing the algorithm. Moreover, axons in real neurons propagate signals in the forward direction only.

In the following we will discuss learning algorithms which are more similar to the processes which take place in biological neurons. These algorithms are based on the back-propagation algorithm.

The learning algorithms are given for higher order networks [3], [7]. Such networks (see e.g., [8]) are able to solve much more complicated problems than those solved by neural networks of the first order. The description of the algorithms is given without loss of generality for the particular case when a multilayer network consists of neurons of the second order.

7.2.2 Neural Networks With a Single Output

First, let us discuss a single-output neural network with non-negative memory elements $P = \left\{ w_i^{(\mu,v)} \leq 0, w_{kl}^{(\mu,v)} \leq 0 \ \forall \ k,l,v; \ i = \overline{1, N^{(\mu,v)}}; \ \mu = \overline{2, N_M} \right\}$ and non-negative continuously increasing activation functions

$\psi^{(\mu,\nu)} \geq 0 \ \forall \nu, \mu = \overline{1, M-1}$. These limitations mean that all threshold weights $w_0^{(\mu,\nu)}$, all weights in the first layer, and outputs of the last layer may have any signs; all other weights and all other outputs must be non-negative. Under such limitations on the memory and the activation functions, the elements of the transient sensitivity matrix $D^{(i+1)}$ (7.13) are non-negative and, hence, in view of equation (7.27), the signs of all components of the sensitivity vector $S^{(i)}$ coincide with the sign of the MNN output error. It is clear that this conclusion holds true independently of the network order.

In virtue of the assumptions mentioned above the expression for the gradient

$$\nabla_{W^{(\mu)}}Q = u^{(\mu)}Diag(\psi'(net^{(\mu)}))S^{(\mu)T}$$

turns out to be equal to $\nabla_{W^{(\mu,\nu)}}Q = \rho^{(\mu,\nu)}\varepsilon^{(1)}u^{(\mu,\nu)}$ for any neuron, where $\rho^{(\mu,\nu)} \geq 0$. This expression coincides with that of an isolated neuron (or with that of the last layer). Therefore, it is possible to conclude that the adaptation of the neural network memory with the constraints mentioned above is carried out as if a network error was given to every separate neuron, i.e., as it presumably happens in natural neural networks.

In the case under consideration if, for example, $Q(\varepsilon(w;m)) = \varepsilon^T(w;m)\varepsilon(w;m)/2$, the network learning procedure can be written for every separate neuron independently:

$$w^{(\mu,\nu)}(n) = P_Q\Big[w^{(\mu,\nu)}(n-1) - \gamma_P \varepsilon^{(1)}\big(w(n-1)\big)u^{(\mu,\nu)}(n)\Big], \qquad (7.31)$$

where $\gamma_P > 0$ is a constant characterising the length of the correction step, P_Q is the projection operator onto the set P of non-negative memory elements.

It is easy to show that the (7.31) is a pseudo-gradient training procedure with respect to function J (7.3) or (7.4), and under some conditions it provides a solution in a certain neighbourhood of the conditional minimum of the function J. We shall denote this learning procedure as an "*autonomous learning algorithm with constraints*" because every neuron learns independently (in some sense), employing the network error and its own extended input vector $u^{(\mu,\nu)}(n)$. Note, that algorithm (7.31) does not depend on the nature of connections between the network layers, and in this sense it is a structurally independent algorithm. Nevertheless one can expect that the efficiency of algorithm (7.31) will vary with the structure. Thus, it is essential to search for the most efficient structure such that the conditional minimum determined by algorithm (7.31) is close to the unconditional minimum.

One could expect that the increase of the network power (the number of neurons, number of layers, dimension of the input vector, etc.) would imply that the domain of the unconditional optimal values of the memory includes the set Q. In this case the solution provided by the autonomous learning algorithm with constraints (7.31) should coincide with the solution given by the back-propagation method, which was described in the first part of the chapter.

7.2.3 Neural Network With Many Outputs

Consider a network with many outputs under the conditions of Section 7.2.2. In this case the gradient of the function under minimisation turns out to be equal to

$$\nabla_{w^{(\mu,v)}} Q = \left(\sum_{i=1}^{N_M} \varepsilon^{(i)} \rho^{(\mu,v,i)} \right) u^{(\mu,v)}, \tag{7.32}$$

where $\rho^{(\mu,v,i)} \geq 0$ for any neuron. Note that $\rho^{(\mu,v,i)} \equiv 0$, if the i-th output of the network is not connected with the output of the (μ,v)-neuron. If the gradient (7.32) is substituted by its estimate $\overline{\nabla_{w^{(\mu,v)}} Q} = \left(\sum_{i=1}^{N_M} \varepsilon^{(i)} \lambda \right) u^{(\mu,v)}$ on the set Q by choosing the best (in some sense) positive value of λ, then we shall obtain an algorithm of the following form:

$$w^{(\mu,v)}(n) = P_Q \left[w^{(\mu,v)}(n-1) - \gamma_p \overline{\varepsilon} \left(w(n-1) \right) u^{(\mu,v)}(n) \right], \tag{7.33}$$

where

$$\overline{\varepsilon_n} = \begin{cases} \displaystyle\sum_{j=1}^{N_M} \varepsilon_n^{(j)} & \text{- if the information about the connection of } (\mu,v)\text{-neuron to } i\text{-th output of the network is ignored,} \\[2em] \displaystyle\sum_{j} \varepsilon_n^{(j)} & \text{- in the opposite case with summation being carried out only over those numbers } j \text{ of output neurons which are connected to the } (\mu,v)- \text{ neuron.} \end{cases} \tag{7.34}$$

Algorithm (7.33)-(7.34) will be called the "*autonomous algorithm with constraints for a multiple output network*".

It should be noted that the autonomous property of the back-propagation algorithm holds also when there are no sign constrains on the thresholds weights $w_0^{(\mu,\nu)}$ $\forall\ \mu,\nu$. This follows directly from Eqs. (7.21) and (7.27).

The autonomous algorithms proposed in [9] are simple for computer implementation, do not depend on the structure of the multilayer neural network and are closer to their neurobiological analogues than the usual algorithms. These properties make them a very convenient device for the training of neural networks, especially when the network consists of a large number of neurons. Such algorithms can be effectively implemented in transputer networks or in neurochips to design on-line pattern recognition systems of large dimensions.

References

[1] B. T. Polyak, Introduction to optimization. New York: Optimization Software, Inc, Publications Division, 1987.
[2] Ya. Z. Tsypkin, Adaptation and learning in automatic systems. New York: Academic Press, 1971.
[3] D. E. Rumelhart, G. E. Hinton, and R. J. Williams, "Learning internal representations by error propagation", Parallel Distributed Processing, V.1, ch. 8, pp. 675-695, D. E. Rumelhart and J. L. McClelland, Eds., Cambridge, MA: MIT Press, 1986.
[4] K. Narendra, and K. Parthasarathy, "Identification and control of dynamic systems using neural networks", IEEE Trans. on NN, No. 1, pp. 4-27, 1990.
[5] L. Fortuna, S. Geaziani, M. L. Presti, and G. Muscato, "Improving back-propagation learning using auxiliary neural networks", Int. J. Contr., v.55, No. 4, pp. 793-807, 1992
[6] Y. Hirose, K. Yamashita, and S. Hijiya, "Back-propagation algorithm which varies the number of hidden units", Neural Networks, v.4, pp. 61-66. 1991
[7] Y. C. Lee, G. Doolen, H. H. Chen, G. Z. Sun, T. Maxwell, H. Y. Lee, and C. L. Giles, "Machine learning using a higher order correlation network", Physica D, v. 22, pp. 276-306, 1986.
[8] P. Peretto, J. J. Niez, "Long term memory storage capacity of multi-connected neural networks", Biol. Cybern., v. 54, pp. 53-63, 1986.
[9] E. D. Avedyan, M. L. Kovalenko, L. E. Tsitolovsky, and Y. Z. Tsypkin, "Autonomous learning algorithms for neural networks", Proceedings of the International Conference Mathematics, Computer, and Investments" pp. 2-11, 15-19 Feb.1993, Moscow, Garant, 1993.

8 Identification and Control of Dynamic Systems Using Multilayer Neural Networks

8.1 Identification of Dynamic Systems Using Multilayer Neural Networks

The identification problem can be stated as follows:

Let $u_P(n)$ and $y_p(n)$ be the input and output signals of a time-invariant dynamic plant D_p. The plant is assumed to have known parameterisation but with unknown values of the parameters. The objective is to construct a suitable identification model D_M which produces an output $y_M(n)$ close to $y_p(n)$ in some sense, when subjected to the same input $u_P(n)$ as the plant.

8.1.1. Linear Systems

Let us discuss the linear time-invariant discrete-time case. The plant equation can be written as a linear difference equation of the N-th order

$$y_p(n) = \sum_{i=1}^{N} a_i y_p(n-i) + \sum_{j=0}^{M} b_j u_p(n-d-j), \qquad (8.1)$$

where a_i and b_j are constant unknown parameters and $d \geq 1$ is a positive integer characterising the plant's delay. The following two identification models with adjustable parameters $\overline{a}_i$ and $\overline{b}_j$ are possible:

$$y_M(n) = \sum_{i=1}^{N} \overline{a}_i y_M(n-i) + \sum_{j=0}^{M} \overline{b}_j u_p(n-d-j) \qquad (8.2)$$

(dynamic, or parallel model) and

$$y_M(n) = \sum_{i=1}^{N} \overline{a}_i y_p(n-i) + \sum_{j=0}^{M} \overline{b}_j u_p(n-d-j) \qquad (8.3)$$

(static, or series-parallel model).

The series-parallel model is preferable Since it is a combiner and there are many algorithms for estimation of the parameters of a linear combiner, such as the projection algorithms, the least squares algorithms and so on.

8.1.2 Non-Linear Systems

A non-linear plant can be described with the following non-linear difference equation of the N-th order:

$$y_p(n) = \Psi\left(y_p(n-1), ..., y_p(n-N), u_p(n-d), ..., u_p(n-d-M); a\right), \quad (8.4)$$

where a is the vector of the parameters of the non-linear system (8.4).

If we know the structure of the plant but not the vector of the parameters a, then a suitably parameterised identification model will be the series-parallel static model described by the equation

$$y_M(n) = \Psi\left(y_p(n-1), ..., y_p(n-N), u_p(n-d), ..., u_p(n-d-M); \overline{a}\right). \quad (8.5)$$

From the comparison of (8.4) and (8.5) it follows that the vector of the parameters $\overline{a}$ in the model (8.5) exist (for example, $\overline{a} = a$), so that both plant and model have the same output for any specified input. Hence, the identification procedure consists in adjusting the parameters $\overline{a}$ in the model (8.5) using the optimisation methods described in Chapter 3 based on the error between the plant output (teacher's instruction) and the model output.

If we do not know the structure of the plant (8.4) or the structure is very complicated then we can use some approximation of the non-linear function

$\Psi(\cdot)$. As stated in the previous chapter, good approximation of any continuous function can be achieved with multilayer neural networks. An obvious approach for non-linear system identification is then to choose the structure of the multilayer neural network to yield a good approximation of the right hand side of equation (8.4). Denoting the output of the network as $y_M(n)$ we then have

$$y_M(n) = NN\left(y_p(n-1),...,y_p(n-N),u_p(n-d),...,u_p(n-d-M)\right), \quad (8.6)$$

where $NN(\cdot)$ represents the non-linear mapping of the network. Now the identification procedure consists in adjusting the parameters of the neural network using, for example, the back-propagation algorithm. In this case the teacher's instruction is the output $y_p(n)$ of the plant to be identified.

8.1.3 The Structure of MNNs for Identification

The question about the optimal structure (the number of the layers, the number of the neurons in each layer, the structure of the connections and so on) of the MNN for the identification problem has yet to be solved. The optimal structure problem is avoided by using redundant MNNs. Notice that this implies a smaller rate of convergence. In [1] some simulation results of non-linear plant identification are presented. For relatively simple non-linear functions of two or three variables three layer MNNs containing approximately 30 neurons are used, and approximately 50,000 memory correction steps necessary for convergence.

To update the network weights in [2] the recursive prediction error learning algorithm, which is a Newton type algorithm, is used. This algorithm has much better convergence properties compared with the back-propagation algorithm. The possibility to use the Newton type algorithm is based here on the fact that non-linear functions can be approximated with only one hidden layer neural networks. This structure is based on the results of Kolmogorov [3], Cybenko [4] and Funahashi [5], (see Chapter 6), from which it follows that provided certain conditions hold, a two layer MNN can approximate any continuous function with desired accuracy. The activation function of the single neuron of the second (last layer) is chosen to be linear, whereas neurons of the hidden layers have non-linear activation function.

With these restrictions on the structure MNN the output of the network is given by

$$y^{(2,1)} = w_0^{(2,1)} + \sum_{v=1}^{N_1} w_v^{(2,1)}\psi(w_0^{(1,v)} + w^{T(1,v)}x\,) =$$

$$= w_0^{(2,1)} + \sum_{v=1}^{N_1} w_v^{(2,1)} \psi(w^{T(1,v)} u^{(1)}) = w^{(2,1)T} u^{(2)} ,\qquad(8.7)$$

where x is an input vector for the neural network, $u^{(1)} = (1; x_1, x_2, \ldots, x_{N_x})$ is an extended input vector for the neurons of the first layer, $u^{(2)} = (1, \psi(net^{(1,1)}), \psi(net^{(1,2)}), \ldots, \psi(net^{(1,N_1)}))^T$ is an extended input vector for the neuron of the second layer, $net^{(1,v)} = w^{T(1,v)} u^{(1)}$, and $w^{(1,v)}, w^{(2,1)}$ are the memory vectors of the neurons of the first layer and the neuron of the second layer, respectively.

The instantaneous performance index in [2] has the form

$$Q(\varepsilon(n)) = \frac{1}{2}\varepsilon^2(n) ,\qquad(8.8)$$

where

$$\varepsilon(n) = y^{(*,1)}(n) - y^{(2,1)}(n)\qquad(8.9)$$

is the error signal at the output neuron and $y^{(*,1)}(n)$ is the teacher's instruction.

The instantaneous gradient and Hessian corresponding to performance index (8.8) can be written in the form

$$\nabla_w Q(\varepsilon(w;n)) = -\varepsilon(w;n)\nabla_w y^{(2,1)}(n) ,\qquad(8.10)$$

and

$$\nabla_w^2 Q(\varepsilon(w;n)) = \nabla_w\left(\nabla_w^T Q(\varepsilon(w;n))\right) = \nabla_w\left(-\varepsilon(w;n)\nabla_w^T y^{(2,1)}(n)\right) =$$
$$= -\nabla_w y^{(2,1)}(n)\nabla_w^T y^{(2,1)}(n) - \varepsilon(w;n)\nabla_w^2 y^{(2,1)}(n) .\qquad(8.11)$$

It is important to emphasise however that in [2] only the first member $-\nabla_w y^{(2,1)}(n)\nabla_w^T y^{(2,1)}(n)$ is used for the Hessian. This is possible because of the small influence of the second member on the Hessian (in the linear case the second member is equal to zero).

The gradient $\nabla_w y^{(2,1)}(n)$ and the second gradient $\nabla_w^2 y^{(2,1)}(n)$ can be easily computed using expression (8.7) for function $y^{(2,1)}(n)$.

Expressions (8.10) and (8.11) allow us to write Newton's algorithm (see Chapters 3 and 4) in the following form, if only the first member in the Hessian is taken into account:

$$w(n) = w(n-1) - \left(\sum_{m=1}^{n} \nabla_w^2 Q(\varepsilon(w(n-1);m)) \right)^{-1} \nabla_w Q(\varepsilon(w(n-1);n)) = \quad (8.12)$$

$$= w(n-1) + \Gamma(n)\varepsilon(w(n-1);n)\nabla_w y^{(2,1)}(n),$$

where matrix

$$\Gamma(w(n-1)) = \left(\sum_{m=1}^{n} \nabla_w^2 Q(\varepsilon(w(n-1);m)) \right)^{-1} \quad (8.13)$$

can be calculated recursively (see Chapter 5):

$$\Gamma(w(n)) = \Gamma(w(n-1)) - \frac{\Gamma(w(n-1))\nabla_w y^{(2,1)}(n)\nabla_w^T y^{(2,1)}(n)\Gamma(w(n-1))}{1 + \nabla_w^T y^{(2,1)}(n)\Gamma(w(n-1))\nabla_w y^{(2,1)}(n)} .(8.14)$$

In [2] a forgetting factor $\lambda(n)$ is used for calculating $\Gamma(w(n))$:

$$\Gamma(w(n)) = \frac{1}{\lambda(n)}\left(\Gamma(w(n-1)) - \frac{\Gamma(w(n-1))\nabla_w y^{(2,1)}(n)\nabla_w^T y^{(2,1)}(n)\Gamma(w(n-1))}{\lambda(n) + \nabla_w^T y^{(2,1)}(n)\Gamma(w(n-1))\nabla_w y^{(2,1)}(n)} \right). \quad (8.15)$$

This factor allows the correction of non-exact expressions for the Hessian.

The derivation, implementation and properties of this algorithm are reported in [6]. This algorithm has superior convergence properties when compared with the back-propagation algorithm.

8.2 Control of Dynamic Systems Using Multilayer Neural Networks

Here we will only discuss the problem of the application of MNNs to model reference adaptive control (MRAC) systems. The MRAC problem can be qualitatively stated as follows:

A plant with input-output pair $\{u(n), y_p(n)\}$ and with known structure is given. The plant parameters are unknown. A stable reference model M is specified by its input-output pair $\{r(n), y_M(n)\}$. The aim is to determine the control input $u(n)$ for all n so that

$$\lim_{n \to \infty} \left| y_p(n) - y_M(n) \right| \le \varepsilon$$

for some specified constant $\varepsilon \ge 0$.

8.2.1 Linear Systems

The following linear difference equation of the N-th order describes a SISO plant:

$$y_p(n) = \sum_{i=1}^{N} a_i y_p(n-i) + \sum_{j=0}^{M} b_j u_p(n-d-j), \tag{8.16}$$

where a_i and b_j are constant unknown parameters and $d \ge 1$ is a positive integer characterising the plant's delay. To find the control law $u(n)$ let us transform the right side of Eq. (8.16) by repeated substitution of the expression $y_p(n)$. We have

$$y_p(n) = \sum_{i=1}^{N} \tilde{a}_i y_p(n-d-i+1) + \sum_{j=0}^{M+d-1} \tilde{b}_j u_p(n-d-j). \tag{8.17}$$

Assuming that $y_P(n) = y_M(n)$ and changing n-d for n in Eq.(8.17), we can find the control function $u(n)$ from the following equation:

$$y_M(n+d) = \sum_{i=1}^{N} \tilde{a}_i y_p(n-i+1) + \sum_{j=0}^{M+d-1} \tilde{b}_j u_p(n-j) \tag{8.18}$$

or

$$u_p(n) = \frac{1}{b_0} \left(y_M(n+d) - \sum_{i=1}^{N} \tilde{a}_i y_p(n-i+1) - \sum_{j=1}^{M+d-1} \tilde{b}_j u_p(n-j) \right). \tag{8.19}$$

If control law (8.19) is stable and parameters $\tilde{a}_i, \tilde{b}_j$ are known, then the plant output $y_P(n)$ (8.17) will track the reference signal $y_M(n)$. If parameters

$\tilde{a}_i, \tilde{b}_j$ are unknown, it is possible to estimate them using equation (8.18) and a projection algorithm, such as the least squares algorithm, and substitute these estimates into the control law (8.19).

It should be noted that control law (8.19) utilises an inverse system of the plant given by Eq. (8.17). In fact, Eq. (8.19) represents the inverse model of the plant in which   the variable $y_P(n+d)$ is replaced by the reference signal $y_M(n+d)$.

8.3 Control of Non-Linear Dynamic Systems Using Multilayer Neural Networks

A non-linear plant can be described with the following non-linear difference equation of the N-th order:

$$y_p(n) = \Psi\Big(y_p(n-1),...,y_p(n-N),u_p(n-d),...,u_p(n-d-M);a\Big). \quad (8.20)$$

To use the approach described above for the linear case we transform the right side of Eq.(8.20) by repeated substitution of expression $y_p(n)$, so that

$$y_p(n) = \tilde{\Psi}\begin{pmatrix} y_p(n-d),...,y_p(n-d-N+1), \\ u_p(n-d),...,u_p(n-2d-M+1);a \end{pmatrix}, \quad (8.21)$$

where $\tilde{\Psi}(\cdot)$ is a new non-linear function.

The inverse model construction can be accomplished now using MNNs. For this purpose let the MNN input vector be of the form

$$x(n) = \begin{pmatrix} y_p(n), y_p(n-d),...,y_p(n-d-N+1); \\ u_p(n-d-1),...,u_p(n-2d-M+1) \end{pmatrix}$$

Denoting the output of the network as $u_{NN}(n)$ we then have

$$u_{NN}(n-d) =$$
$$NN\begin{pmatrix} y_P(n), y_p(n-d),..,y_p(n-d-N+1); \\ u_p(n-d-1),...,u_p(n-2d-M+1) \end{pmatrix} \quad (8.22)$$

In Eq. (8.22) $NN(\cdot)$ represents the non-linear mapping of the network's input vector $x(n)$. Now the learning procedure consists of adjusting the parameters of the neural network using, for example, the back-propagation algorithm. The teacher's instruction is in this case the control signal $u_p(n-d)$.

The inverse model (8.22) inserted into the control loop will be described now by the following non-linear recursive equation:

$$u_P(n) = NN\left(\begin{array}{l} y_M(n+d), y_p(n), .., y_p(n-N+1); \\ u_p(n-1), .., u_p(n-d-M+1) \end{array}\right). \tag{8.23}$$

A necessary condition for the implementation of the control law (8.23) is that it is stable. In this case the output of the non-linear plant will follow the reference signal.

A detailed survey of neural networks from a control system perspective can be found in [7].

8.3.1 Example use of Neural Networks for MRAC

An example architecture for MRAC of non-linear plants using neural networks is shown in figure 8.1. The role of the network is to form an inverse model of the plant to be controlled. During training, the network inputs are delayed values of the plant input and output. The desired output of the network is the current plant input signal. Obviously training will only be successful if an inverse model exists for the plant and that model is unique.

During operation, the network is used to calculate the current input, $u(n)$, which will cause the plant output at time $n+d$ to be equal to that of the model. Thus if a unique plant inverse model exists and the resulting controller architecture is stable, then an effective control system can be developed.

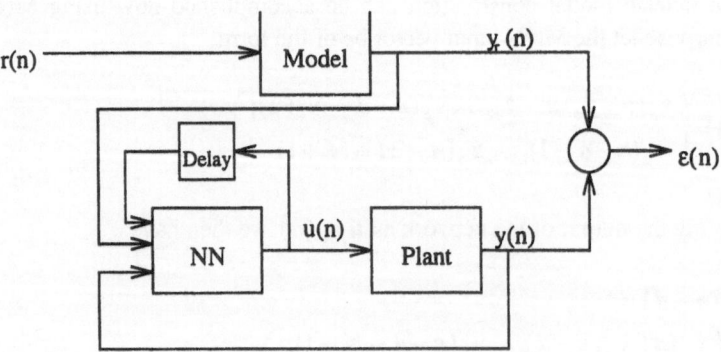

Fig 8.1. Example showing Neural Networks for model reference adaptive control of a non-linear plant.

References

[1] K. Narendra, and K. Parthasarathy, "Identification and control of dynamic systems using neural networks, "*IEEE Tr. on NN* ", No. 1, pp. 4-27, 1990.

[2] S. A. Billings, H. B. Jamaluddin and S. Chen, "Properties of neural networks with applications to modelling non-linear dynamic systems", Int. J. Control, vol. 55, No. 1, pp. 193-224, 1992.

[3] A. N. Kolmogorov, "On the representation of continuous functions of several variables by superpositions of continuous functions of one variable and addition", Doklady Akademii Nauk SSSR, vol. 114, No. 5, pp. 953-956, 1957.

[4] G. Cybenko, "Approximation by superpositions of a sigmoidal function", Mathematics of Control, Signals and Systems, 2, pp 303-313, 1989.

[5] K. Funahashi, "On the approximate realization of continuous mappings by neural networks", Neural Networks, 2, pp 183-192, 1989.

[6]. A. Billings, H. B. Jamaluddin and S. Chen, "A comparison of the back-propagation and recursive prediction error algorithm for training neural networks", Mechanical System and Signal Processing, 5, 233-255, 1991.

[7] K. J. Hunt, D. Sbarbaro, R. Zbikowski and P. J. Gawthrop, "Neural networks for control systems-a survey", Automatica, vol. 28, No. 6, pp. 1083-1112, 1992.

9 The Cerebellar Model Articulation Controller (CMAC)

9.1 Introduction to CMAC

9.1.1 Introduction

In 1975 Albus [1], [2] presented a new approach to the solution of the robot control problem by means of a controller he called a CMAC (Cerebellar Model Articulation Controller). An application of this approach may be found, for example, in [3]. Albus' approach was effectively implemented by Ersü and Millitzer in 1984 (see, for example [4], [5]), and they named their system AMS (Associate Memory System). AMS was implemented in the LERNAS adaptive system for control of dynamic objects [4].

One of the basic elements of CMAC-type adaptive control systems is the associative memory AMS/CMAC, which stores information on the input-output relationships of a non-linear dynamic system. Learning in a CMAC is performed by means of an algorithm presented in [2] by Albus. The convergence properties of the learning algorithm for the CMAC were studied in detail in [6], [7].

An important parameter of CMAC is the learning process convergence rate. This parameter determines the possibility of controlling plants with time-varying parameters, and the faster the convergence rate for training, the more non-stationary objects can be controlled by means of the CMAC. The learning process convergence rate can be increased in an CMAC with fixed parameters (memory size, number of active memory cells, coding method, etc.) by changing the learning algorithm.

We will discuss here a modified CMAC-algorithm [11] that has a higher convergence rate than the algorithm presented by Albus [2], concentrating on the properties of the proposed algorithm.

We will also discuss some aspects related to the application of the CMAC to identification and control of non-linear dynamic plants.

9.2 Data Storage and Learning Process in the CMAC

9.2.1 Introduction to the CMAC

The CMAC is intended for effective storage, restoration, and interpolation of multidimensional functions. The operating principles of the CMAC can be briefly stated as follows:

An input vector s (containing the values of the arguments of the function to be stored) is coded by a special mapping algorithm so that it is associated with a fixed number ρ of active associative memory cells in the AMS, where the total number of cells is p and $p \gg \rho$. The unique feature of the CMAC is the mapping algorithm which transforms the distance between input vectors s_1 and s_2 into the degree of overlap between sets of addresses A_1 and A_2 where the functional values are stored. The result is that the number of elements in the intersection A_1 and A_2 is roughly proportional to the closeness in the input-space of two input vectors s_1 and s_2 regardless of the dimensionality of the input.

The relationship between the input vector s and the active memory cells is uniquely described by a p-dimensional association vector a in which each element $a_i, i = 1, ..., p$ may take the value zero or one. Zeros corresponds to inactive associative cells, and ones are placed in positions corresponding to active associative cells. We will denote the one and zero-elements of the association vector as the active and inactive elements of this vector, respectively. The associative memory cell of index i contains a number w_i called a weight. The weights w_i form a p-dimensional weight vector w. The elements of the association vector a and the weight vector w are ordered in the same way according to the indices of the CMAC memory cells. The weights w_i are chosen so that the average value of the active weights is equal to the value of the function to be stored. This method of organising the memory in a CMAC (in which information on the values of a stored function is stored in the form of cells of associative memory) allows interpolation of the values of a stored function, i.e., the possibility of automatically predicting the value of a function that has not yet been stored in CMAC.

We must distinguish two different processes in the CMAC: the training process, in which an algorithm determines the weights w_i using values of the function and its arguments, and a second process, the restoration process, in which an input vector s (the arguments of the function) is used to restore or estimate the values of the function r. During the training process, when the CMAC in the k-th cycle is presented with input vector $s(k)$ and a corresponding function value $r(k)$, the weight vector w is corrected so that it takes the value $w(k)$ at the k-th step. The training process is constructed so that as the number of observations k increases, the sequence $w(k)$ approaches a stationary value.

A trained CMAC makes it possible to restore or estimate the value of a function from its arguments in accordance with a specific rule: the input vector s uniquely determines an association vector a, which, in turn, determines the active memory cells whose arithmetic average is the estimated value of the function.

9.2.2 The CMAC system

In order to calculate the p-dimensional association vector a it is necessary to know the subscripts of the active cells.

One-Dimensional Case: Each real input variable s_i in the input vector $S = \left(s_1, s_2, ..., s_N\right)^T$ is first converted into a binary variable m_i. The following rule is used for this transformation:

1. Each digit in the binary variable m_i has a value of "1" over one and only one interval within the range of s_i and is equal "0" elsewhere.
2. The number of bits set to "1" in each binary variable m_i is constant and equal to ρ, usually $\rho=2^k$, and k is an integer.
3. The subscripts of the binary digits in m_i^* are tabulated against the value of the variable s_i.

Multidimensional Case: The mapping in this case consists of N individual mappings $s_i \rightarrow m_i^*$ for all the variables in the input vector $S = \left(s_1, s_2, ..., s_N\right)^T$ and the concatenation of the corresponding elements in each of the m_i^*.

Mapping Into a Memory of Practical Size: It is possible to have an additional mapping $A \rightarrow A_p$ such that the association cells in a very large set A are mapped onto a much smaller set A_p. One way in which this can be done is the hash-coding. Hash-coding operates by taking the address of where an item of data is to be stored in the large memory and using it as an argument in a pseudorandom number generator for computing an address in the smaller memory.

The details can be found in [1].

9.3 Albus' Learning Algorithm

An algorithm for the training CMAC was proposed by Albus in [2]. This algorithm is recursive, so the correction $\Delta w(k)$ for the weight vector $w(k\text{-}1)$ based on $k\text{-}1$ measurements is computed according to the following rule: when the k-th value of the input vector $s(k)$ and the corresponding value of the function $r(k)$ are received, a coding algorithm is used to compute the p-dimensional association vector $a(k)$ from the input vector $s(k)$, and, consequently, active and inactive associative elements of the vector $w(k)$ are determined, the arithmetic average of the active weights of the vector $w(k\text{-}1)$ is subtracted from the value of the function $r(k)$ stored; the obtained difference is added to the values of the active weights; the values of the inactive weights are not changed. Albus' training algorithm can be written analytically in the form

$$w_i(k) = w_i(k-1) + \left(r(k) - \sum_j w_j(k-1)/\rho \right), \left(\forall\, i, a_i(k) = 1;\ \forall\, j, a_j(k) = 1 \right), \quad (9.1)$$

$$w_i(k) = w_i(k-1), \left(\forall\, i, a_i(k) = 0 \right). \quad (9.2)$$

The sequence $w(k)$ in (9.1)-(9.2) is constructed so that the arithmetic average of the corrected active weights is equal to the stored value of the function $r(k)$ in the k-th step, i.e.,

$$r(k) = \sum_i w_i(k)/\rho\ , \left(\forall\, i,\ a_i(k) = 1 \right).$$

This condition follows directly from Eq. (9.1).

The convergence properties of the sequence $w(k)$ in (9.1)-(9.2) were investigated by Militzer and Parks [6]. They showed that, depending on the coding rule, which provides an association vector for each input vector s, the sequence $w(k)$ converges either to a point or to a region, which they called the zone of attraction.

In [6] the proof of convergence for the sequence $w(k)$ is based on a representation of algorithm (9.1)-(9.2) in the equivalent vector form

$$w(k) = w(k-1) + \frac{(\rho\, r(k) - a^T(k)w(k-1))}{a^T(k)a(k)} a(k), \quad (9.3)$$

where $w(k)$ and $a(k)$ are column vectors and T indicates transposition.

Despite the fact that relation (9.3) does not contain the normalising factor $\rho^{1/2}$ that appears in [6], representation (9.3) is equivalent to the representation of [6].

We should note that Albus' algorithm in the form (9.3) is the same as the known Kaczmarz algorithm [8]-[10] for the iterative solution of the system of linear equations.

$$a^T(k)w = \rho\, r(k), \; k = \overline{1, N}, \tag{9.4}$$

in which, in contrast to Albus' algorithm, the elements of vector $a(k)$ are arbitrary real numbers, whereas here they are binary numbers. Thus, Albus' algorithm is a special case of Kaczmarz's, so any conclusions concerning the latter apply to the former.

9.4 Modified Albus' Algorithm

9.4.1 Modification

The Kaczmarz algorithm has been used by specialists in automatic control theory as an effective means for solving identification and adaptive control problems with small measurement errors and, more important, slow drift in the parameters of the observed or controlled object.

One of the most important characteristics of the Kaczmarz algorithm is its convergence rate, which determines the number of iterations required for the training algorithm to solve a system (9.4) with a given accuracy. The convergence rate for the algorithm depends closely on the properties of the input sequence $a(k)$. In particular, if the sequence $a(k)$ is a sequence of orthogonal vectors, the solution w^* can be found in a finite number p of iterations, where p is the dimension of vector w. In other cases, the convergence rate of Kaczmarz algorithm may prove to be inadequate for determining the way in which solutions w^* of system (9.4) vary in the non-stationary case. This makes it necessary to increase the convergence rate of the algorithm.

The convergence rate of the algorithm can be increased by modifying the way in which each point $w(k)$ is computed through an orthogonal projection of the preceding point $w(k-1)$ onto the k-th hyperplane. In algorithms using this modification, the point $w(k)$ is computed by orthogonal projection of the point $w(k-1)$ onto the intersection of the preceding l hyperplanes $a^T(n)w = \rho\, r(n), \; n = k, k-1, ..., k-l+1$, where the parameter l, in turn, may be a function of the iteration number k. The parameter l can be thought of as the memory depth of the algorithm (see Chapter 2).

In [11] the following modification of the Kaczmarz algorithm was proposed to increase the convergence rate for the learning process in CMAC:

$$w(k) = w(k-1) + b(k)(\rho r(k) - a^T(k)w(k-1)), \tag{9.5}$$

where

$$b(k) = A_l(k)\left(A_l^T(k)A_l(k)\right)^{-1}e \tag{9.6}$$

is the correction vector,

$$A_l(k) = (a(k), a(k-1), \ldots a(k-l+1)) \tag{9.7}$$

is $p \cdot l$ - block matrix, and

$$e = (1, 0, \ldots 0)^T . \tag{9.8}$$

Indeed, by (9.7) and (9.8), we can represent vector $b(k)$ of (9.5) and (9.6) in the form

$$\dot{b}(k) = \beta_0 a(k) + \beta_1 a(k-1) + \ldots + \beta_{l-1} a(k-l+1), \tag{9.9}$$

where the coefficients $\beta_0(k), \beta_1(k), \ldots, \beta_{l-1}(k)$ are the elements of the first column of the $l \cdot l$-matrix $\left(A_l^T(k)A_l(k)\right)^{-1}$. For small values of l the coefficients $\beta_i(k)$, $i = 0, \ldots, l-1$, can be computed as functions of the elements of $a(n)$, $n = k, k-1, \ldots, k-l+1$; for large values of l they can be computed by inversion of the matrix

$$A_l^T(k)A_l(k) = \begin{bmatrix} \alpha_{k,k} & \alpha_{k,k-1} & \cdot & \alpha_{k,k-l+1} \\ \alpha_{k-1,k} & \alpha_{k-1,k-1} & \cdot & \alpha_{k-1,k-l+1} \\ \cdot & \cdot & \cdot & \cdot \\ \alpha_{k-l+1,k} & \cdot & \cdot & \alpha_{k-l+1,k-l+1} \end{bmatrix}$$

where $\alpha_{i,j} = a^T(i)a(j)$ is the number of ones in common positions of the vectors $a(i)$ and $a(j)$, $\alpha_{i,j} \leq \alpha_{i,i} = \alpha_{j,j} \leq \rho$, or by solving the corresponding system of equations.

The elements $\alpha_{i,j}$ of the matrix $A_l^T(k)A_l(k)$ and the elements of vector $b(k)$ can be computed rather simply without great computational power. First, the p-dimensional vector $a(k)$ is uniquely described by a ρ-dimensional vector whose elements are the indices of the elements in the vector $a(k)$ whose values are one. Second, because of the binary properties of the vector $a(k)$, the elements of the p-dimensional vector $b(k)$ in (9.9) can have only 2^l different values at any step k.

For the case $l=2$, the vector $b(k)$ can be written as:

$$b(k) = \frac{\rho}{\rho^2 - \alpha^2_{k,k-1}} a(k) - \frac{\alpha_{k,k-1}}{\rho^2 - \alpha^2_{k,k-1}} a(k-1)$$

so that the modified Albus algorithm takes the form

$$w_i(k) = w_i(k-1) + \frac{1}{1 + \tilde{\alpha}_{k,k-1}}\left(r(k) - \sum_j w_j(k)/\rho\right) \quad (9.10)$$

$$\left(\forall\, i,\, a_i(k) = 1,\, a_i(k-1) = 1;\ \forall\, j, a_j = 1\right),$$

$$w_i(k) = w_i(k-1) + \frac{1}{1 - \tilde{\alpha}^2_{k,k-1}}\left(r(k) - \sum_j w_j(k)/\rho\right) \quad (9.11)$$

$$\left(\forall\, i,\, a_i(k) = 1,\, a_i(k-1) = 0;\ \forall\, j, a_j = 1\right),$$

$$w_i(k) = w_i(k-1) + \frac{-\tilde{\alpha}_{k,k-1}}{1 - \tilde{\alpha}^2_{k,k-1}}\left(r(k) - \sum_j w_j(k)/\rho\right) \quad (9.12)$$

$$\left(\forall\, i, a_i(k) = 0, a_i(k-1) = 1;\ \forall\, j, a_j = 1\right),$$

$$w_i(k) = w_i(k-1) \quad (9.13)$$

$$\left(\forall\, i,\, a_i(k) = 0,\, a_i(k-1) = 0\right)$$

where $\tilde{\alpha}_{k,k-1} = \alpha_{k,k-1} / \rho$ is the proportion of unit elements in the vectors $a(k)$ and $a(k-1)$.

There are several special characteristics of algorithm (9.10)-(9.13) with $l=2$ that are characteristic of the modified algorithm for arbitrary $l \leq p.$.

If the input vector $a(k)$ is orthogonal to the preceding vector $a(k-1)$, then $\alpha_{k,k-1} = \tilde{\alpha}_{k,k-1} = 0$ and Eqs. (9.10) and (9.11) become Eq. (9.1), while Eqs. (9.12) and (9.13) become (9.2), i.e., in this case the modified Albus algorithm becomes the ordinary Albus algorithm.

If an input vector $a(k)$ is parallel to the vector $a(k-1)$, then $\tilde{\alpha}_{k,k-1} = \alpha_{k,k-1} / \rho = \rho / \rho = 1$ and the expression in the denominator of relations (9.10)-(9.12) becomes equal to zero. In this case the computation is impossible.

9.4.2 Special Features of the Modified Albus Algorithm

The special features of the modified Albus algorithm are due to its construction: an estimate $w(k)$ is computed by orthogonal projection of the preceding estimate $w(k-1)$ onto the intersection of l hyperplanes. This projection can be performed only when all l preceding vectors $a(n)$, $n = k, k-1, ..., k-l+1$ are linearly independent. In this case, the non-negative definite Gram matrix $A_l^T(k)A_l(k)$ is positive definite and is invertible, since the correction vector $b(k)$ can be computed with formula (9.10).

1. If, however, $a(k)$ is linearly dependent on the $l-1$ vectors $a(n)$, $n = k-1, ..., k-l+1$, the determinant of the Gram matrix is equal to zero and there is no solution of the form (9.10). Actually, the measurements $a(k)$ and $r(k)$ carry no additional information on the solution w^* compared to the preceding $l-1$ measurements. As a result, when the Gram matrix degenerates, the corresponding k-th measurement must be rejected and the process must be continued with the $(k+1)$-th measurement as the k-th.
2. The convergence rate of the modified algorithm increases as the memory depth l of the algorithm increases.
3. If system (9.4) is consistent, as, for example, in the case of additive measurement noise at the output variable $r(k)$, the estimates computed with both Albus and the modified Albus algorithm converge to some trap zone. The size of such a zone increases as the memory depth l increases in the modified algorithm, i.e., the faster the algorithm, the more sensitive it is to noise in measurements.

4. The algorithm is less sensitive to measurement noise when, during the measurement of $a(k)$, the Gram matrix $A_l^T(k)A_l(k)$ is ill-conditioned. The degree of definiteness in the Gram matrix can be characterised by the value of its determinant, which determines the threshold below which measurements cannot be used to compute $w(k)$.
5. In general, the effectiveness of the modified algorithm is clear when the input sequence is subject to slow changes. This occurs, for example, in a cyclic training process, when the input sequence is generated by scanning the entire input space. Here, as a rule, $a_{ij} \neq 0$.

If, however, the input points are far from each other during the training process, the modified algorithm provides no advantages over Albus algorithm. This occurs, for example, in a purely random training process, when any given point is selected totally independently of the preceding point.

Results of numerical comparisons between Albus algorithm and the modified algorithm can be found in [11].

9.5 CMAC for Identification and Adaptive Control

From the first part of the Chapter it follows that the CMAC can be effectively used for storage, restoration, and interpolation of multidimensional functions. The approximation capabilities of CMAC are discussed in [12]. Thus, it is possible to apply the CMAC to identification and adaptive control in the same manner as MNNs were used in the previous chapter. A comparison between CMAC neural network control and two traditional adaptive control systems can be found in [13].

References

[1] Albus J. S. "A new approach to manipulator control: the cerebellar model articulation controller", ASME Trans., J. Dynamic Systems, Measurement ant Control, 97, No. 3, pp. 220-227, 1975.

[2] Albus J. S. "Data storage in the cerebellar model articulation controller (CMAC)", ASME Trans., J. Dynamic Systems, Measurement ant Control, **97**, No. 3, pp. 228-233, 1975.

[3] Miller W. T., Glanz F. H., Kraft L. G. "CMAC: An associative neural network alternative to backpropagation", Proceedings of the IEEE, vol. 78, No. 10, pp. 1561-1567.

[4] Ersü E., Tolle H. "A new concept for learning control inspired by brain theory", in: Proc. Ninth IFAC World Congress, vol. 2, Budapest, pp. 1039-1044, 1984.

[5] Tolle H., Militzer J., Ersü E., "Zur Leistungsfähigheit lokal verallgemeinernder assoziative Speiher und ihren Einsatzmöglichkeiten in lernenden Regelungen", Messen, Steuern, Regeln, 3, pp. 98-105, 1989, (in German).

[6] Militzer J., Parks P. C., "Convergence properties of associative memory in trainable control systems", Automation and Remote Control, No. 2, 1989.

[7] Parks P. C., Militzer J., , "Convergence properties of associative memory storage for learning control systems", in: IFAC Symposium on Adaptive Systems in Control and Signal Processing, vol. 2, Glasgow, UK, pp. 565-572, 1989.

[8] Kaczmarz S., "Àngenährte Auflösung von Systemen linearer Gleichungen", Bulletin International de l'Academie Polonaise des Sciences. Lett A, pp. 355-357, 1937

[9] Kaczmarz S., "Approximate solution of systems of linear equations", Int. J. Control, vol. 57, No. 6, pp. 1269-1271, 1993

[10] Parks P. C., "S. Kaczmarz", Int. J. Control, vol. 57, No. 6, pp. 1263-1267, 1993

[11] Aved'yan E.D., Hormel M., "Increasing the convergence speed for the training process in special associative memory system", Automation and remote control, No.12, pp. 1723-1730, 1991.

[12] Cotter N. E., Guillerm T. J.," The CMAC and a theorem of Kolmogorov", Neural Networks, vol. 5, No. 2, pp. 221-228, 1992.

[13] Kraft L. G., Campagna D. P., "A comparison between CMAC neural network control and two traditional adaptive control systems", IEEE Control System Magazine, Special issue on neural networks in control systems, 10, No. pp, 36-43, 1990.

Index